CROWN AND COMPANY

CROWN AND COMPANY

THE
HISTORICAL RECORDS
OF THE
2nd BATT. ROYAL DUBLIN FUSILIERS

FORMERLY THE
1st BOMBAY EUROPEAN REGIMENT

BY

COLONEL H. C. WYLLY, C.B.

The Naval & Military Press Ltd

Published by
The Naval & Military Press Ltd
5 Riverside, Brambleside, Bellbrook
Industrial Estate, Uckfield, East Sussex,
TN22 1QQ England
Tel: +44 (0) 1825 749494
Fax: +44 (0) 1825 765701
www.naval-military-press.com

[*Photo, Gale & Polden, Ltd.*

Dedicated

TO

FIELD-MARSHAL

HIS ROYAL HIGHNESS ARTHUR W. P. A.

THE DUKE OF CONNAUGHT AND STRATHEARN,

K.G., K.T., K.P., G.C.B., G.C.S.I., G.C.M.G., G.C.I.E., G.C.V.O., G.B.E.,

COLONEL-IN-CHIEF

THE ROYAL DUBLIN FUSILIERS,

IN GRATEFUL RECOGNITION

OF HIS UNFAILING INTEREST

IN HIS BATTALION

OF "OLD TOUGHS."

MAJOR-GENERAL C. D. COOPER, C.B.,
COLONEL THE ROYAL DUBLIN FUSILIERS.

FOREWORD

I believe it is generally accepted that Ireland's losses in the Great War were proportionally heavier than those of her sisters in the United Kingdom and that despite the indignity of exclusion from participating in the national call for conscription, with all its dire consequences ; thus, owing to our large casualties, it has been by no means easy for the historian to compile the second volume of Crown and Company, and his difficulties have been augmented by the disbandment of the Battalion. I think, however, all ranks of the " Old Toughs " will agree that Colonel Wylly, C.B., has, notwithstanding, carried out his task most admirably. Our best thanks are due, in addition, to Mr. C. S. Seager, Director of the Publishers, for his valuable advice and assistance so readily placed at our disposal.

G. S. HIGGINSON,
Lieut.-Colonel,
late Comdg. 2nd Bn. The Royal Dublin Fusiliers
(" Old Toughs ").

KILDARE STREET CLUB,
DUBLIN.
July, 1923.

CONTENTS

LIST OF ILLUSTRATIONS

LIST OF MAPS

CROWN AND COMPANY

CHAPTER I

1911—1914.

THE EVE AND OPENING OF THE WAR.

THE early history of the 2nd Battalion The Royal Dublin Fusiliers, which, under the title " Crown and Company " was written by Major Arthur Mainwaring and published in 1911, brought the record down to July 1st of that year, on which day the Battalion, then stationed at Aldershot, received new Colours at the hands of His Royal Highness Field-Marshal the Duke of Connaught, Colonel-in-Chief of the Regiment.

Two days later the Battalion proceeded to the neighbourhood of Frensham for brigade training, and early in August divisional training was commenced about Oxney Farm. The summer of this year was abnormally hot, 100° Fahrenheit being registered in the shade, and the operations being protracted and the country shadeless all ranks suffered much from the effects of the sun. The troops returned to Aldershot on August 10th, *ostensibly* owing to the trying weather and the danger of heath-fires necessitating the curtailment of the operations ; but the Battalion had hardly marched into barracks when it became apparent that there were other reasons than those stated for the sudden cessation of the divisional training.

Early in this year there had been a great deal of labour unrest, and this had lately revived, continuing throughout July both in the capital and in Liverpool, Manchester and other great towns. The seamen had been successful in a conflict they had provoked with the Shipping Federation, and this had encouraged other bodies of workers to act likewise, while in certain towns there had been

" strikes in sympathy." On August 1st the men employed at the London docks struck, some twenty ocean liners being held up, while more than 20,000 men were out. This particular strike was settled within ten days, but no sooner had the dockers returned to work than the carmen came out and there was some rioting and disorder, while discontent was now perceptible on nearly all the great railway systems in the country, traffic was greatly interfered with and large quantities of perishable provisions were spoilt. The authorities had seen how matters were drifting and extensive preparations had been made for sending some 20,000 troops to London from Aldershot, Woolwich and elsewhere.

In consequence of the above, at 2.25 p.m. on August 10th orders were received for the 2nd Battalion Royal Dublin Fusiliers to hold itself in readiness to proceed to London that evening for strike duty, and by 4 o'clock all ranks were ready to leave, but the order was cancelled later in the day, and as matters now presented a more settled appearance officers and other ranks were, on the 12th, permitted to go away on the summer leave usually granted at the close of the training season.

On the 14th orders were suddenly issued to recall all those who had proceeded on leave of absence, and although this order was later rescinded it was not in time to stop from returning many individuals who had already left Aldershot. These were of course permitted to go away again only, however, to be finally recalled again on the 16th, and then on the evening of the day following orders were issued which affected practically all the troops in the Aldershot Command, and throughout the night of August 17th-18th corps were entraining for London at the various railway stations in and adjacent to Aldershot.

The orders for the Battalion were, to leave Farnborough Station at 3 a.m. on the 18th, but on arrival there it was found that other troops had not yet completed their entraining, so that it was 3.45 a.m. before the Battalion actually started. On arrival at Waterloo the Brigade was assembled and marched to Regent's Park where it was encamped, and when Londoners began to go about their business in the morning, they discovered greatly to their surprise that the metropolis was full of soldiers. In the afternoon " C," " D," " F " and " G " Companies were sent to guard the termini at King's Cross and St. Pancras, while during the night " E " Company under Major Loveband proceeded to Golder's Green Station for the same duty.

On the 19th some officers and men joined headquarters, when the strength of the Battalion was 26 officers, 1 warrant officer and 583 non-commissioned officers and men ; the following were the officers with the Battalion :—Lieutenant-Colonel W. Bromilow ; Majors A. E. Mainwaring, R. A. Rooth and A. Loveband ; Captains (Brevet Major) W. J. Venour, D.S.O., A. T. Magan, L. F. Renny, K. C. Weldon and J. P. B. Robinson ; Lieutenants E. F. E. Seymour, A. J. D. Preston, R. M. Watson, J. E. Vernon, J. F. K. Dobbs, T. J. Leahy, H. M. Floyd ; Second-Lieutenants E. R. L. Corballis, J. A. C. Hogan, C. H. L'E. West, A. F. Dobbs, E. R. L. Maunsell, F. S. Lanigan-O'Keeffe, W. H. Braddell and A. G. Astley ; Captain and Adjutant H. W. Higginson and Captain and Quartermaster J. Burke.

Early on the afternoon of the 19th it was announced that the Brigade was to move and while the destination was not stated, rumour was busy as usual ; all detachments were called in and the Brigade left the camp in the Regent's Park at 5.15 p.m., the units separating and the Royal Dublin Fusiliers marching to St. Pancras where, owing to the total disorganization induced by the strike, entraining was a matter of considerable difficulty, and it was 7.30 at night before the Battalion steamed off into the unknown. It was not until arrival at Leicester that a staff officer met the train and handed the commanding officer instructions for the disposal of the companies as follows :—

Headquarters and 6 companies—Kirkby.

1 company—Westhouses.

1 company—Pyebridge.

But just as the train was about to move off again Colonel Bromilow was summoned to the telephone where he was directed by General

Graham, the Brigadier, to take the Battalion to Claycross, dropping two companies *en route* at Ambergate.

On arrival at Ambergate " A " and " H " Companies (5 officers and 126 other ranks under Captain Magan) were left there, and the remainder steamed on to Claycross, which was reached at 2 a.m. on the 20th. Here news was received of serious rioting at Chesterfield, and a telephone message was handed in from the Mayor of that town begging that the troops might be sent on thither immediately on arrival at Claycross, and this, the approval of the Brigadier having been obtained, was accordingly done. The Headquarters, with six companies of the Battalion arrived at Chesterfield at 4 a.m. to find that every pane of glass in the station building had been broken by the rioters, who had only dispersed on hearing of the approach of the soldiers. At Chesterfield the Battalion relieved two companies of infantry on duty at the railway station—one furnished by the West Yorkshire Regiment and the other by the Gordon Highlanders.

On the 21st the Battalion supplied two detachments, Captain Robinson and 20 men proceeding to Hasland, where trouble was thought to be brewing among the men of the Locomotive Department, while Brevet Major Venour with 3 officers and 69 other ranks were sent to Staveley; everything, however, was quiet and the attitude of the crowds towards the soldiers was perfectly friendly.

The situation at this time was far more serious than the majority of the people of this country had any idea by reason of the critical condition of Anglo-German relations, due to what was known as " the Agadir incident," a condition, indeed, of which the public had no cognizance until the following November; but deliberations as to the labour questions under dispute had now for some time past been going on until, late on the night of August 19th, a settlement was reached; the strike was to be ended, the men's leaders were to persuade the men to go back to work, all strikers were to be reinstated and in no way penalized, while conciliation boards were to meet at once and settle all questions in dispute.

On the 23rd and 25th the two outlying detachments were

called in to Headquarters and finally on August 27th the Battalion
returned by train from Chesterfield to Aldershot.

Things here now followed their usual course and the Aldershot
Command Rifle Meeting, the progress of which had been interrupted
by the strikes, was resumed on the return of the troops to their
stations. At the meeting the 2nd Battalion Royal Dublin Fusiliers
was very successful, winning the " Young Soldiers' Cup " with a
score of 361 and being second in the " Aldershot Command Challenge
Cup " and in the " Wellington Cup " ; in the company competitions
" F " Company did particularly well, while many individual prizes
were won by non-commissioned officers and men of the Battalion.

The 2nd Battalion Royal Dublin Fusiliers was now under orders
to leave Aldershot for Gravesend, but prior to its departure the
Battalion team, in charge of Captain Robinson and Lieutenant
Maunsell, carried off the " Obstacle Challenge Shield " presented
by Field-Marshal the Duke of Connaught, defeating the second team
by 20 points. The following telegram was received by the Commanding Officer from His Royal Highness :—

*" Duke of Connaught congratulates all ranks on having won his
Obstacle Challenge Shield and is very glad it should have been won by
his own Regiment."*

On September 27th the Battalion left Aldershot by train and
arrived the same day at Gravesend, there to be quartered. It had
been preceded by an advance party composed of Captain Robinson,
Captain and Quartermaster Burke and 16 non-commissioned
officers and men, and the officers quartered with the Battalion at
its new station were :—Lieutenant-Colonel Bromilow, Majors
Mainwaring, Rooth and Loveband, Brevet Major Venour, Captain
and Adjutant Higginson, Captains Magan, Renny, Smith, Tredennick and Robinson, Lieutenants Seymour, Trigona, Watson,
Vernon, Dobbs and Leahy, Second-Lieutenants Corballis, Hogan,
West, Dobbs, Maunsell, Luke, Priest and Braddell, with Captain
and Quartermaster Burke. The strength of the Battalion in
" other ranks " was 2 warrant officers, and 814 non-commissioned
officers and men with 38 women and 44 children.

The Battalion was now in the 10th Infantry Brigade, Eastern Command. The Brigade was commanded by Brigadier-General the Hon. E. Stuart-Wortley, C.B., C.M.G., D.S.O., M.V.O., while Lieutenant-General Sir A. H. Paget, K.C.B., K.C.V.O., A.D.C., was G.O.C.-in-C. of the Command.

From Gravesend the Fusiliers furnished two detachments, one, composed of 7 officers—Brevet Major Venour, Lieutenant Vernon, Second-Lieutenants Corballis, Hogan, Luke, Maunsell and Braddell—and 250 non-commissioned officers and men, was sent to Lodge Hill, while the other proceeded to Sheerness for duty; with this latter were Captain Smith, Second-Lieutenant West and 50 other ranks.

When in the previous July H.R.H. the Duke of Connaught had presented the Battalion with new Colours, he had been offered, and had been graciously pleased to accept, those which had been replaced; and on October 3rd these, carried by Colour-Sergeants Hall and McDonagh, were handed over to the Colonel-in-Chief by Lieutenant-Colonel Bromilow.

This was the last official act connected with his command of the Battalion by Colonel Bromilow, for on March 5th, 1912, his period of service in command came to an end and he was succeeded by Major A. E. Mainwaring.

In the first half of this year there was a recrudescence of the strike trouble of the previous year, and on June 10th the Battalion was ordered to " stand by " by reason of unrest produced by a strike among the London dockmen. Later in the same month, as a result of strike disturbances at Tilbury, 400 men of the Royal Dublin Fusiliers were sent thither from Gravesend to protect the railway line between Tilbury and Grays stations on the London, Tilbury and Southend Railway. At one time matters assumed an ugly look, for a crowd of some 2,000 strikers armed with cudgels assembled at the level crossing at Grays, but they fortunately contented themselves with " booing " and shaking their fists at the " Old Toughs."

By August 3rd the trouble had all subsided and such detach-

GROUP OF OFFICERS, 1912.

Back Row (left to right)—Lt. T. J. Leahy, Capt. N. P. Clarke, Lt. J. E. Vernon, Lt. W. F. Higginson, Lt. C. D. Priest, Lt. E. F. F. Seymour, Lt. E. R. L. Maunsell, Lt. J. Luke, Lt. A. S. Trigona, Lt. A. Dobbs, Lt. W. H. Braddell.
Sitting—Capt. J. P. Tredennick, Lt. R. M. Watson, Capt. J. Burke, Major A. E. Mainwaring, Lt.-Col. W. Bromilow, Major R. A. Rooth, Major A. Loveband, Capt. H. M. Shewan, D.S.O., Capt. E. St. G. Smith.

ments as had been employed on strike duty were withdrawn, when a letter, of which the following is an extract, was addressed to the Commanding Officer by the Chief Constable of Essex :—

" I should like to take this opportunity of thanking you and the officers and men of the Royal Dublin Fusiliers, on behalf of the County of Essex and myself, for the invaluable services you all have rendered us under most trying circumstances ; and I feel sure that had it not been for their presence very serious disturbances would have arisen with probably great destruction and loss of property."

During the second week in August the Battalion—strength 24 officers and 414 other ranks—left Gravesend and went into camp at Ash, near Aldershot, for brigade and divisional training ; this was followed, from September 15th to the 20th, by Army manœuvres in East Anglia, and at the end of the training period the following orders were promulgated :—

By the Brigadier-General :—

" At the conclusion of Collective Training the Brigadier-General Commanding desires to express his appreciation of the splendid spirit displayed by all ranks throughout the above period, and more especially when camped at Aldershot and during the Manœuvres. The Brigade has attained a high standard of efficiency, and it is a particular source of satisfaction for the Brigadier-General Commanding to know that through the spirit of comradeship which pervades all ranks, the 10th Infantry Brigade is really one unit and not merely a collection of four battalions. The Brigadier-General Commanding wishes to thank all ranks for this excellent result, which, without their ready and willing co-operation, could not have been attained."

By the Director of Manœuvres :—

" The King has desired the Director of Manœuvres to inform the troops of the great pleasure it has given him to pass the last three days among his troops during the final phase of their annual training, and to witness the zeal, keenness and energy displayed by both officers and men under service conditions in marching and fighting.

" His Majesty, from his inspections of the Army during the past

two and a half years, feels satisfied that the present system of training is being conducted on sound lines and is particularly pleased to observe that a serious and earnest effort to reach a high standard of war efficiency prevails.

" The first experiments in aerial reconnaissance—which have caused His Majesty and the country to mourn the loss of four gallant officers; the rapid concentration of troops by rail without the dislocation of ordinary traffic; and the employment of mechanical transport are special features of the Manœuvres of 1912.

" The King desires to express his entire satisfaction with everything he has seen of the work in the field, both on the part of the troops and of the staff."

During the course of the month following the close of the training the Battalion was re-armed with the Lee-Enfield Rifle Mark III; and on October 21st the 2nd Royal Dublin Fusiliers furnished a Guard of Honour to H.H. Princess Marie Louise of Schleswig-Holstein, on her paying a visit to Gravesend, whither the Battalion had returned from Aldershot; and at Gravesend it remained quietly for the next few months pursuing the usual duties of peace soldiering.

In August 1913 the Battalion left Gravesend for Shorncliffe, where this year brigade training was carried out, and at the close of this it was entrained for Wolverton—strength 21 officers and 533 other ranks—and was engaged in divisional training and Army exercises until September 28th when the Battalion returned to Gravesend.

In March 1914 Army Order 279 of the year previous was given effect to: this order contained the following laconic amendment to the " Dress Regulations " so far as these concerned the officers of the Royal Dublin Fusiliers: " Badges. On the Service Dress. On the Collar in bronze; as for forage caps, in pairs, but without scroll."

This order resulted from suggestions made by Lieutenant-Colonel A. E. Mainwaring, commanding the 2nd Battalion, who

TALANA GROUP, 1913.

Back Row (left to right)—Pte. M'Donnell, L.-Cpl. Codger, Pte. Smith, Pte. Toole, Pte. Holmes, Pte. Ryan, Pte. Brady, Pte. Adby, Pte. Kelly.

Second Row—Pte. Bolton, Pte. Reilly, Pte. Maher, Pte. Mackey, Pte. Carroll, Pte. Murray, Pte. Whelan, Pte. Andrews.

Front Row—C.S.M. Williams, Sergt. Hayes, Colour-Sergt. Hatt, Colour-Sergt. Richards, Major Shewan, D.S.O., Capt. and Quartermaster Burke, Colour-Sergt. Sutton, Sergt.-Dmr. Davies, Sergt. Harper, Sergt. Grace, Sergt Smith.

had represented that were the existing scroll cut off, leaving only the grenade, a very much neater and more suitable collar badge would result. The matter was referred to the officers of the Regiment, who were unanimously in favour of the alteration, which was further approved by H.R.H. Field-Marshal the Duke of Connaught, Colonel-in-Chief, and by Major-General C. D. Cooper, Colonel, the Royal Dublin Fusiliers.

During the early part of the year 1914 the internal troubles, which now came to a head in Ireland owing to the controversy raised by the Home Rule Bill, caused the nation generally to disregard the far more serious or indeed mortal dangers which on the Continent were commencing to threaten the peace of the world. In March rumours reached London that many Army officers serving with regiments at the Curragh, particularly those of the Third Cavalry Brigade there stationed, had resigned or offered to resign their commissions in order to avoid serving against Ulster should any attempt be initiated by the Cabinet to coerce the loyalists of that province.

The trouble in Ireland, and certain symptoms of unrest in many other parts of the United Kingdom, were not allayed for some considerable time ; but suddenly in July all domestic difficulties were to a great extent obscured and even removed by the gathering of dense war clouds in Central Europe. In this month the news became public of the murder on June 28th at Serajevo of the heir to the throne of Austria, but few men, in this country at least, realized the effect which the crime might have upon the peace of the civilized world. The Government of the Dual Monarchy, however, presented a strongly-worded ultimatum to the Serbian Cabinet, and by July 26th it became known that Austria-Hungary had rejected with contumely the very reasonable reply which had been received from Belgrade.

Before the situation had become in any way actually critical, the British Foreign Minister, Sir Edward Grey, had made proposals for the convening of a conference of Ambassadors to discuss the matters at issue, the military operations which the Vienna Government was already projecting being in the meantime suspended ; Germany, however. on her part, refused attendance at the proposed conference, and thus encouraged, aided and abetted, Austria-Hungary at once declared war upon Serbia. In reply to this Russia on July 28th mobilized fourteen of her Army Corps.

By August 1st the European situation had become worse ; German troops were preparing to invade France ; Russia had proclaimed a general mobilization ; Germany had announced her immediate intention of acting in like manner ; and in Great Britain steps were taken for the safeguarding of all magazines, dockyards, tunnels and bridges and for the defending of our harbours by means of booms ; while from the remotest parts of the British Empire, where men had realized the danger that seemed to be threatening the Homeland, came cable after cable containing offers of material help and assurances of moral support.

On August 2nd the Naval Reserves were called out ; the newspapers of the following day announced Germany's violation of the neutrality of Luxemburg, her invasion of France and her insolent ultimatum to Belgium ; on the 4th was issued the Royal Proclamation calling out our Army Reserve and decreeing the embodiment of the Territorial Force ; and on the same day the Court of St. James' demanded of the German Government an " unequivocal assurance that Germany would respect the neutrality of Belgium, identical with that given the week before by France both to Belgium and to Great Britain, and for a satisfactory reply by midnight," otherwise the British Ambassador in Berlin was instructed to make what was virtually a declaration of war.

Late that night this request was refused, and on Wednesday morning, August 5th, Great Britain found herself at war with Germany and the Powers then or thereafter allied with her.

On this day a Council of War was held at No. 10 Downing Street, and was attended by nearly all the members of the Cabinet and by Field-Marshal Sir John French, the Commander-in-Chief elect of the British Expeditionary Force, when the following matters were discussed :

[Photo, W. T. Munns, Gravesend.

"C" COY., 2ND BN. THE ROYAL DUBLIN FUSILIERS, 1914,

(*a*) The Composition of the Expeditionary Force.
(*b*) Its point of concentration in France.

It was generally believed that it had been mutually agreed between the British and the French military authorities that, in the event of it becoming necessary for our troops to render material assistance to those of France, we should send a division of cavalry and six divisions of infantry across the Channel; but this arrangement was of course liable to be modified according to the number of troops which our government might consider it necessary to retain in the United Kingdom to guard against the possibility of invasion and to meet the forces of disorder. After some considerable discussion it was decided that two divisions of all arms must for the present remain in England, and that the rest of the Expeditionary Force, of a strength of roughly 100,000 men, should be sent to France with the utmost dispatch possible.

The point of concentration was then considered, but was not at that session of the War Council finally settled. Hitherto it had been agreed between the British and the French General Staffs that the Expeditionary Force should operate on the left flank of the French armies, and that the actual detraining area should be between Maubeuge and Le Cateau. Lord Kitchener, who was now appointed Secretary of State for War, held the opinion that this concentration position was too far forward and suggested that the neighbourhood of Amiens was preferable. Events seem to show that his views were the sounder, but the arrangement originally made was adhered to.

The Expeditionary Force then which left England in the early part of August 1914 consisted of two army corps, each of two divisions, and a cavalry division. The Ist Army Corps contained the 1st and 2nd Divisions and was commanded by Lieutenant-General Sir Douglas Haig, while the IInd was made up of the 3rd and 5th Divisions and was in the first instance placed in charge of Lieutenant-General Sir James Grierson, who, on his death in the train very soon after landing in France, was succeeded by General

Sir Horace Smith-Dorrien.   The commander of the cavalry division was Major-General Allenby.

The 4th Division was thus temporarily detained in England for purposes of home defence, but it was understood that it would later follow the Expeditionary Force to the Continent—indeed to all thinking men it was abundantly clear that every man and gun that could be trained and raised " would be required, in conjunction with our Allies, to assist in repelling the rolling tide of German invasion."*

In August 1914 the 4th Division, now commanded by Major-General T. D'O. Snow, C.B., was composed and distributed as follows :—

10th Infantry Brigade, Brigadier-General J. A. L. Haldane, C.B., D.S.O.   1st Battalion Royal Warwickshire Regiment, 2nd Battalion Seaforth Highlanders, 1st Battalion Royal Irish Fusiliers, all at Shorncliffe, and 2nd Battalion Royal Dublin Fusiliers at Gravesend.

11th Infantry Brigade, Brigadier-General A. G. Hunter-Weston, C.B., D.S.O.—1st Battalion Somerset Light Infantry, 1st Battalion East Lancashire Regiment, 1st Battalion Hampshire Regiment, and 1st Battalion Rifle Brigade, at Colchester.

12th Infantry Brigade, Brigadier-General H. F. M. Wilson, C.B. —1st Battalion Royal Lancaster Regiment, 2nd Battalion Lancashire Fusiliers, and 2nd Battalion Royal Inniskilling Fusiliers—all at Dover, and 2nd Battalion Essex Regiment at Chatham.

The 4th Division later formed part of the IIIrd Army Corps under Lieutenant-General Pulteney.

Already on July 27th certain measures had been taken in the Gravesend area for safeguarding places of importance, and Captain Trigona and 28 men were sent to Port Victoria, while Lieutenant Leahy and 24 non-commissioned officers and men proceeded to Thameshaven.

Two days later—on Wednesday, 29th—a cypher telegram was received directing that " Precautionary Period Stations " should be taken up and the following detachments were sent out :—

* Haldane, *A Brigade of the Old Army*, p. 1.

GROUP OF OFFICERS, 1913.

Back Row (left to right)—Lt. T. J. Leahy, Lt. F. C. S. Macky, Capt. R. L. H. Conlan, Lt. J. MacN. Dickie, Lt. B. McGuire, Lt. H. A. Shadforth, Lt. J. E. Vernon, Lt. C. H. L'E West, Lt W. H. Braddell, Capt. A. S. Trigona.

Sitting—A. Walker Esq., Capt. N. P. Clarke, Lt. R. M. Watson, Major H. M. Shewan, D.S.O., Lt.-Col. A. E. Mainwaring, Capt. J. Burke, Capt. W. H. Supple, Capt. G. S. Higginson, Capt. G. N. Cory, D.S.O.

Two Platoons of " D " Company to Chatham.

" A " and " B " Companies and two Platoons of " D " Company to the Isle of Grain.

" C " Company to Coalhouse Fort, Tilbury Docks and Thames-haven Oilworks.

On the 30th an officer and 25 men were sent to Cliffe Fort.

At 5.5 p.m. on August 4th the order to mobilize was published, while on the evening of the 7th, before even the mobilization was completed, a telegram was received containing orders for the 10th Brigade and a battery to move by rail to York, where they would come under the orders of Sir Herbert Plumer, the General Commanding-in-Chief the Northern Command.

At midnight instructions reached the 2nd Battalion Royal Dublin Fusiliers that their train would leave in half an hour, but it was not until 12.30 a.m. on the 8th that the Battalion—strength 20 officers and 805 other ranks—actually marched to the railway station, Lieutenant Leahy and a party remaining behind to bring on reservists and horses which had not yet joined. Owing to the absence of transport animals, all the wagons had to be man-handled to the station.

The Battalion left Gravesend in two trains at 3.45 and 5.15, and reached York at 11.30 a.m. and 1 p.m., marching then about a mile to the Knavesmire Race Course, where the Battalion and two companies of the Royal Warwickshire Regiment were accommodated in the Grand Stand, the remainder of the Brigade being billeted in York or in the store sheds of the North-Eastern Railway Company.

During the next few days there was a good deal of route marching to get the feet of the reservists hardened ; men were inoculated against enteric, the transport horses arrived from Gravesend, and the Brigade was inspected by General Plumer.

On August 18th the 10th Brigade left York by train for Harrow, where the 4th Division was now concentrating, and on the evening of the 20th it was announced that the Division was to depart in the small hours of Saturday, the 22nd, for " an unknown destination."

On that day the Battalion left Harrow in two trains at 3.5 and 4.5 a.m., and reached Southampton at 7 and 8 o'clock in the morning.

" Little time was lost in getting the troops and transport on board the two large steamers which were to carry the 10th Infantry Brigade across the Channel. My headquarters, the 1st Battalion Royal Warwickshire Regiment, the 2nd Battalion Royal Dublin Fusiliers, and a battery of the 29th Field Artillery Brigade took possession of the s.s. *Caledonia* of the Anchor Line, which sailed under sealed orders at 11.30 a.m.   The day was fine and sunny."*

The 2nd Battalion The Royal Dublin Fusiliers, " the Old Toughs," had now set out upon the greatest adventure in the long and distinguished career of the Regiment, and few can have foreseen how long the war was to last, and—happily, perhaps—none of that joyous band of adventurers realized how few of them would return from it.

The Battalion embarked at a strength of 22 officers, 6 warrant officers, 57 sergeants and lance-sergeants, 70 corporals and lance-corporals, 9 drummers, and 881 privates.

The following were the officers who left England with the Battalion :—Lieutenant-Colonel A. E. Mainwaring, Major H. M. Shewan, D.S.O., Captains G. S. Higginson, N. P. Clarke, R. L. H. Conlan, S. G. de C. Wheeler, W. H. Supple, A. S. Trigona, and R. M. Watson (Adjutant) ; Lieutenants J. E. Vernon, J. F. K. Dobbs, T. J. Leahy, C. H. L'E. West, and W. H. Braddell ; Second-Lieutenants J. MacN. Dickie, F. C. S. Macky, B. Maguire, J. G. M. Dunlop, R. A. J. Goff, and W. M. Robinson ; Captain and Quartermaster J. Burke, and Captain P. C. T. Davey, R.A.M.C., in medical charge.

Captain A. J. D. Preston and Lieutenant H. A. Shadforth had been sent to the Depot on embarkation of the Battalion to assist in forming the nucleus of the Service Battalions now about to be raised, and both joined the first of these new units—the 6th (Service) Battalion Royal Dublin Fusiliers.

* Haldane, p. 8.

CHAPTER II

FRANCE, 1914.

LE CATEAU—THE MARNE—THE AISNE—ARMENTIÈRES.

" NIGHT had closed in as we neared Boulogne Harbour," wrote
the commander of the 10th Infantry Brigade, " the approach to
which was no longer marked by the fitful beams of the familiar
lighthouse. . . . As we steamed slowly up the entrance to the
port we were greeted by the cheers and ' Vivent les Anglais !' of a
goodly number of the inhabitants, to which the troops replied.
By 10.30 p.m. we were alongside the wharf. Soon the troops began
to disembark, and marched to a camp on the high ground above the
town near the *Colonne de la Grande Armée*, myself and staff remaining
on board till morning. I then heard that the remainder of my
brigade had landed at Havre, whence they did not join me until
the 24th."

Here the troops were at once pestered for " souvenirs," the
inhabitants all appearing desirous of forming a collection of the
cap and shoulder badges of every unit of the British Army. " This
passion for souvenirs," writes General Haldane, " followed us everywhere, and in spite of strict orders against parting with any portion
of their uniform, I should say that a large percentage of the population of Northern France managed to secure a metal memento of
the British soldier. Later on, when we were retreating before the
Germans and passing through a village, I was told that one of the
Dublin Fusiliers in my brigade, who was wearily dragging himself
along in the ranks of his company, hearing the too familiar cry of
' souvenir,' turned an angry glance over his shoulder and growled,
' Here, you can have my blooming pack for a souvenir !' "

" About 3 p.m. on Sunday, the 23rd, reports began coming in to the effect that the enemy was commencing an attack on the Mons position, apparently in some strength, but that the right of the position from Mons and Bray was being particularly threatened."* In the extraordinary way that rumour, especially of an unfavourable character, is spread, it had already become known in Boulogne tolerably early on this day that all was not going any too well at the front. German cavalry, too, was reported to be already south of Ostend, while hostile motor cyclists were stated to have been seen within ten miles of Boulogne itself. There were, however, to the north-east French Territorial troops along the bank of the River Aire, and Brigadier-General Haldane had been instructed that in case of need the two battalions with him, and two more of the 12th Brigade, were to assist in driving off the invaders.

However, the 10th Brigade and its commander were not required to make any stay in the camp above Boulogne ; orders were early issued for a move, and in the evening the 2nd Battalion Royal Dublin Fusiliers left camp and entrained at 9 p.m., but it was not until midnight that the trains started.

Corporal McBirney most unfortunately broke his leg in two places in helping to bring the regimental transport down the hill to the town.

Journeying by way of St. Amiens and St. Quentin, the Battalion reached Le Cateau during the morning of August 24th, and here detrained and marched some four miles along the Cambrai road to the vicinity of the villages of Beaumont and Inchy, south of which the troops bivouacked, the Dublin Fusiliers furnishing the outposts. It had been expected that the Division would almost at once move forward, but later, in consequence, no doubt, of the general obscurity of the situation at the front, orders were issued to stand fast.

Already on the afternoon of the day previous, the 23rd, the Commander of the British Expeditionary Force had received what he describes as " a most unexpected message from General Joffre," the French Generalissimo, telling him that three German corps

---

* Sir John French's despatch of September 7th, 1914.

were moving against his position in front, that another was engaged in a turning movement from the direction of Tournai, and that the French forces on the British right were already in retreat, the Germans having on the day previous seized the passages of the Sambre between Charleroi and Namur.

In consequence of the above, the British forces were obliged also to fall back, and their retirement commenced at daybreak on the 24th, the IInd Corps taking up the line Dour—Quarouble—Frameries, while the Ist Corps reached on the evening of this day the line Bavai—Maubeuge, the IInd Corps conforming and occupying a position to the west of Bavai.

The retirement of both corps was recommenced in the early morning of the 25th to a position in the neighbourhood of Le Cateau, and General Snow, commanding the 4th Division, was ordered to move out to take up a position with his right south of Solesmes, and his left resting on the Cambrai—Le Cateau road south of La Chaprie, where it was hoped that it would be able to render effective aid in the retirement of the Ist and IInd Corps.

The 10th Brigade marched northwards at 2 a.m. on the 25th via Viesly in the direction of St. Python, a village situated on the right of the River Selle, near Solesmes, which was reached at 4.30 a.m. Here a brief halt was made for further orders, and at 6 o'clock gun-fire was heard in the distance. The Brigade then moved off again, marching east, and took up a position to the north of the farm of Fontaine-sur-Tertre ; but later it was withdrawn to some more favourable ground nearer to the farm, where, in conjunction with the remainder of the 4th Division, it was to assist in covering the retreat of the divisions of the IInd Corps.

Here the Brigade came under the fire of enemy guns, while the German cavalry showed considerable boldness, pushing patrols almost up to the farm, near which, as night came on, two Uhlans were brought down in front of the line of the Dublin Fusiliers. These were the first shots fired by the Battalion in the war, and were from men of the picquet under Captain Supple.

Orders now came for the 10th Brigade to retire to Haucourt,

a village lying some eight miles off to the south-east, and during this withdrawal the Brigade formed the rear-guard of the Division, marching by Viesly and Bethencourt.   Haucourt was reached about 6 a.m. on the morning of the 26th, and here Brigadier-General Haldane was met by the Divisional Commander and taken to see the position which he proposed to hold ; General Haldane now learnt that at a conference held in the very early hours of that morning Sir Horace Smith-Dorrien, commanding the IInd Corps, had decided to stand and fight.

Generals Snow and Haldane had barely arrived on the ridge which overlooks the village of Haucourt from the north, when they came under enemy rifle fire and discovered that the German advanced guard, composed chiefly of cavalry, and machine guns carried on motor cars, had surprised the outposts and the troops these were covering, and, taking them at a disadvantage, were driving the British off the greater part of the ridge.   Of the battalions of the 10th Brigade, the Seaforth Highlanders and Royal Irish Fusiliers were in reserve in rear of the village of Haucourt ; the Royal Warwickshire Regiment had become involved in an effort to retake the ridge which the Germans had in part succeeded in occupying ; while of the 2nd Battalion Royal Dublin Fusiliers two companies under Major Shewan were holding the high ground immediately east of the village, the remaining companies being kept back behind Haucourt in support.

The Brigade was here on the extreme left flank of the Allied forces, and the enemy appeared to be trying to work round the left from the direction of Cambrai.

Shortly before 5 p.m. Brigadier-General Haldane received orders to retire, covering the withdrawal of the remainder of the Division, but he had been unable throughout the day to get into communication with, or receive any reports from, the O.C. Royal Dublin Fusiliers ; his mind as to their safety was, however, set at rest by seeing them, apparently, conforming to the rearward movement of the other regiments of the Division.

Of the further operations of the 2nd Battalion Royal Dublin

Fusiliers General Haldane gives the following brief account :—
" From accounts which reached me later it seemed that some remnants " (of the Battalion) " were still holding Haucourt under the command of Major Shewan, Royal Dublin Fusiliers. As night fell the Germans cautiously made their way into that village from the north, whereupon Shewan quietly withdrew his men from it in the opposite direction. After wandering about in the dark for some hours his party seems to have reached the neighbourhood of Ligny, which the Germans had already occupied. What exactly occurred there is doubtful, but after a short, sharp fight some were surrounded and taken, while others, under Captain N. P. Clarke of the same regiment, after several adventurous days and nights, managed to march through the advancing Germans to the coast, whence they were shipped to England, rejoining the Brigade in October. Shewan himself, a gallant and determined officer, was taken prisoner, and his services were consequently lost to the British Army throughout the war. His holding on to Haucourt, however, must have helped to conceal the fact of the withdrawal of the main body, and probably prevented the Germans from pushing on as rapidly as otherwise they might have done."

The events of August 26th—so far as the 2nd Battalion Royal Dublin Fusiliers is concerned—are dismissed in the Battalion diary in something less than half-a-dozen words, and while we may glean something from a report furnished by Captain N. P. Clarke of what happened to some at least of the officers and men of the two companies fighting under Major Shewan, it is very difficult indeed to piece together the details of the action in which the headquarter companies were overwhelmed.

Captain Clarke's report of the action at Haucourt and the account of the subsequent adventures of that officer and his party will be quoted from later.

The general account of the fighting on August 26th which here follows is taken from Vol. I of the Official History.

The orders issued from 4th Division Headquarters at 5 p.m. on August 25th, directed that the division was to march to and take up a position Caudry—Fontaine-au-Pire—Wambaix—knoll just west of Seranvillers, and begin entrenching at daylight on the 26th. The 11th and 12th Brigades were to occupy the line indicated, while the 10th Brigade was to be in reserve at Haucourt. But an order, sent out about an hour and a half later, reduced the front to be held by about three miles, viz., from Fontaine-au-Pire to Wambaix only. The position was attacked soon after 6 a.m. on the 26th and the 12th Brigade was particularly hard pressed, but for at least an hour and a half held its own against a German cavalry division (the 2nd) and two Jäger battalions, backed by artillery and numerous machine guns. About 8.45 a.m., however, the Germans appearing to be moving round the left flank of the advanced line, it was decided to withdraw the 12th Brigade, and, to cover this movement, two companies of the Warwicks of the 10th Brigade were ordered to deliver a counter-attack from Haucourt, and these eased the situation although themselves suffering severely ; the 12th Brigade was then withdrawn to the line Ligny—Esnes.

During these movements the German artillery fire was very heavy, but between 9.30 and 10 a.m. the worst of the enemy attack was past, and the 12th Brigade reformed its line from Ligny through Haucourt to Esnes, already occupied by part of the 10th Brigade, while General Haldane, to secure the left flank of the division, sent the Seaforths to a ridge south and somewhat east of Esnes, and by 11 o'clock firing in this quarter had died down. The 11th Brigade, however, had been by this left rather isolated and had suffered considerably though holding firm. Later it also was brought back and by 4 p.m. the 4th Division was holding the line about Ligny.

About 5 p.m. the brigadiers of the 4th Division were ordered to retreat, the 10th Brigade being detailed as rear-guard ; the volume of German gun-fire had by now greatly increased, the Inniskillings had been forced back to the western fringe of Esnes, and the units of the 10th and 12th Brigades were so intermixed as to make difficult the transmission of orders. Part of the 12th Brigade moved off first soon after 5, halting and facing about on the road between Selvigny and Guillémin ; other of its units and with them the half of the Dublin Fusiliers, started later, and though the German guns smothered the road with shrapnel, the British columns managed to escape any serious damage. The 11th Brigade, and the rest of the 12th, held their positions till 6 or even later, while " of the 10th Infantry Brigade, only the Seaforth Highlanders and the greater part of the Irish Fusiliers were under their brigadier's hand. Half the Warwickshires and a good many of the Dublin Fusiliers were still in Haucourt, and the remainder were dispersed in various directions, some as escort to guns, others in small isolated bodies. As with the rest of Sir Horace Smith-Dorrien's force, the enemy not only did not pursue the 4th Division, but did very little even to embarrass the retreat."

The two supporting companies of the Battalion—" C " Company on the right under Captain Wheeler and the other on the left under Captain Conlan—remained in their position behind the village of Haucourt until late in the afternoon, being for the greater part of the day under a tolerably heavy shell-fire. The situation generally had throughout been very obscure, no orders of any kind had reached the Battalion Commander from either Brigade or Divisional headquarters, and it was believed, not without reason, that the companies in the front line had withdrawn. The Commanding Officer now gave orders for the supporting companies of the Battalion to retire, but this had to be carried out in artillery formation as the German gun-fire was very accurate. The ground retired over was broken and undulating, and with darkness coming on it was impossible for close touch to be maintained, and the two companies perforce broke up into small bodies under officers and other natural or self-elected leaders and marched on, almost mechanically, all through the night, until, on the morning of the 27th, those who had " won through " came together and effected such temporary reorganization as was then practicable at Le Catelet. Colonel Mainwaring, with a party of but little more than the strength of a platoon, eventually found himself in St. Quentin, and in his absence Captain Frankland, from Staff Captain, 10th Infantry Brigade, assumed temporary command of what of the Battalion had by then come together on September 6th.

The numbers of the parties which rejoined will be found on a later page of this chapter.

Of the fate of the two companies under the command of Major Shewan the following details are taken from a report made by Captain N. P. Clarke. As stated by Brigadier-General Haldane he had lost touch during the whole of August 26th with two of the battalions of his brigade, viz., the Warwickshire Regiment and the Royal Dublin Fusiliers, and no orders were consequently received by them as to future movements. At dusk the Germans were seen to be in force on the ridge some 1,500 yards in front of Major Shewan's two outpost companies, and that officer now decided to withdraw from the village of Haucourt as it did not appear to him to be possible to hold it with the small force at his disposal ; but before he had commenced to do so the enemy had forced his way into the village and it became apparent that the Fusiliers' left was in the air.

The party, which now contained men and officers of other regiments besides the Dublin Fusiliers, commenced to retreat southwards by a track which it was hoped would enable it to rejoin the Brigade, but those who were immediately under Major Shewan's orders were delayed and started later than the remainder, owing to Major Shewan insisting on carrying off the machine gun which had been in action on the outskirts of the village. This having been brought in a start was made, and, marching all through the night, the detachment at dawn on the 27th found itself in the village of Ligny-en-Cambresis.

Major Shewan now determined to move south towards Clary and the march was resumed, the advanced guard being under Captain Trigona, and the rear-guard being commanded by Captain G. S. Higginson, the point being under Lieutenant West. The advanced guard was almost at once fired upon, at a range of some 500 yards, from the direction of Montigny, one mile south of Clary, and the whole party was halted, for in the dim light of dawn it was at first

CLARY MILITARY CEMETERY.

MEN OF THE 2ND BN. ROYAL DUBLIN FUSILIERS BURIED
IN GRAVE MARKED X, AUGUST 1914.

OLD DISTILLERY ON LIGNY-CLARY ROAD.

SCENE OF STAND BY ROYAL DUBLIN FUSILIERS AND
ELEMENTS OF OTHER UNITS, AUG. 26TH/27TH, 1914.

thought that the firers were British. The firing stopped and Captain Trigona endeavoured to communicate with the distant troops by signal, and on his stating that the party with him belonged to the Royal Dublin Fusiliers, the troops in Montigny waved their head dresses and shouted, " Dublin Fusiliers, right, come on."

Captain Trigona was not, however, wholly satisfied and now asked that a man should be sent forward to meet him, a request which was in process of being complied with, while Major Shewan, believing that Montigny was held by British troops, was now closing up on the advanced guard. Captain Trigona, however, had now recognized the uniforms of the Montigny party as German and at once extended, but the enemy now opened a heavy rifle and machine-gun fire from the front and right front at a range of no more than 300—400 yards, while another enemy party appeared to be moving behind Major Shewan's left to envelop that flank.

This officer, recognizing that the position was impossible, now ordered a retirement, and the troops fell back towards a farm house some 300 yards in rear, but by this time the casualties were mounting up, Major Shewan, Captain G. S. Higginson and Lieutenant West being all three wounded, and the party appearing to be completely isolated and wholly out of touch with any other British troops.

Captain Clarke now assumed command and, with Lieutenants Vernon and Dobbs, endeavoured to organize the defence of the farm-house, but he soon realized that they must continue to retire as the Germans were out-flanking the building, and a retreat was accordingly effected by alternate rushes to another farm some 600 yards in rear. At each rush the casualties became more and more heavy, for the only cover was that afforded by root crops, and the enemy machine-gun and rifle fire was well sustained. The retreat was carried on to the village of Ligny, where the party joined some more men of the Battalion under Captain Trigona, and was then continued, but on almost every side German forces were met with.

Marching practically across the German communications by way of Haucourt, Cattenières, Carnières, Fressies, Aubigny-au-Bac,

Vitry, Lens, Aubigny and Frevent, Captain Clarke finally reached Abbeville, whence he and his men were sent to Boulogne and so on to England, where all were refitted and eventually rejoined their respective corps.

The party which thus marched to safety through the German lines was composed as under :—

> Royal Dublin Fusiliers, 2 officers and 35 other ranks.
> Gordon Highlanders, 8 other ranks.
> Somerset Light Infantry, 8 other ranks.
> Royal Warwickshire Regiment, 5 other ranks.
> Hampshire Regiment, 5 other ranks
> Royal Irish Fusiliers, 4 other ranks.
> King's Own Regiment, 3 other ranks.
> Royal Munster Fusiliers, 2 other ranks.
> East Lancashire Regiment, 2 other ranks.
> West Riding Regiment, 1 other rank.
> Total : 2 officers and 73 other ranks.

The remnants of the 2nd Battalion Royal Dublin Fusiliers marched on August 27th by Le Catelet to Roisel, where they came up with the rest of the Brigade, and here the Fusiliers were temporarily formed into a composite battalion with what was left of the Royal Warwickshire Regiment. On first taking account of the losses sustained by the Battalion it appeared that 12 officers—two of whom were known to have been wounded—and 531 non-commissioned officers and men were missing ; the officers were Major Shewan, Captains Higginson, Clarke, Conlan, Trigona and Davey, R.A.M.C., Lieutenants Vernon, Dobbs, West and Braddell, Second-Lieutenants Macky and Dunlop. Later, however, more detailed information came to hand, officers and men who had become separated rejoined, and the following distribution figures were then arrived at :—

Headquarters, Captain and Adjutant R. M. Watson and 50 other ranks had marched with the divisional troops, Lieutenant-Colonel Mainwaring having become separated from headquarters about 6.30 p.m. on August 26th. Captain S. G. de C. Wheeler and " C " Company rejoined the Battalion at 5 a.m. on August 27th at Le Catelet.

Company Sergeant-Major R. Hall and 100 other ranks became attached to details of other regiments of the Brigade and rejoined the Battalion on September 5th.

Sergeant-Major F. Treacher with 50 men and 1st Line Transport joined the Battalion on August 30th near Carlepont.

Major H. M. Shewan and about 450 non-commissioned officers and men who had formed the firing line on August 26th and retired independently were missing, but no information as to their whereabouts was to hand.

At Roisel a position was temporarily taken up east of the town with the Seaforth Highlanders on the left ; but at noon on the 27th the line was retired some two miles and a new position occupied. Here the Battalion was on the left of the line with the Highlanders on its right and some French cavalry on the left. There was no sign of the enemy and at 7 p.m. the retreat was resumed. " I shall never forget this night march," wrote Brigadier-General Haldane. " It was the fifth night I had passed without sleep, except for some three hours, and as I found to keep awake on horseback was impossible and would only lead to tumbling off, I was forced to tramp along on foot. . . . We passed through Tertry, Monchy-Lagache, Guizancourt, Croix and Matigny. The arrangements made for this march by the divisional staff officer responsible were admirable. Every side road had been blocked by sending men in advance, so as to preclude the possibility of our losing the proper direction in the dark, and in addition a staff officer handed over to me a fresh guide at each village to which we came. . . .

" At one place where we halted some time before midnight a number of country carts, which had been obtained by requisition, were standing at the roadside and were intended to carry those of the men who were least able to walk the remainder of the distance which we had to cover."

Voyennes was reached at 4 a.m. on the 28th, the River Somme

having just previously been crossed by a stone bridge, and it was here that in the morning the extraordinary order was received which had been issued from G.H.Q. on the 27th at Noyon to the effect that " all ammunition on wagons not absolutely required, and other impedimenta, will be off-loaded and officers and men carried to the full capacity of all transport both horse and mechanical." It is said that this *sauve qui peut* order was passed through the IInd Corps headquarters without its G.O.C.-in-C. seeing it, and that directly he heard of it, realizing it was wholly unnecessary and would only create alarm and despondency, he sent off to stop its being acted upon. He was, however, too late, in the case of the 4th Division, which had already burned some of the officers' kits.

The march was resumed at 9 a.m., the 10th Brigade being the advanced guard of the Division, and on reaching the Somme Canal it left some of its battalions, among which were the Royal Dublin Fusiliers, to guard the bridge north-east of Offoy. At 1 p.m. the Brigade withdrew and followed the Division as rear-guard to Bussy where it arrived at midnight, and " where the men, who had had nothing to eat all day, got food from a motor lorry supply column which came up at a most opportune moment and deposited along the roadside the provender which it carried."

On the 29th it was reported that the enemy was following up in force, and the 3rd and 4th Divisions took up a defensive position, the right of the 10th Infantry Brigade being in touch with the 3rd Division at Crisolles, the left extending to about Cendrière where it communicated with the 12th Brigade at Chévilly. At 9 p.m. the outposts withdrew to another position not far in rear at Beaurains, where the Battalion was in the front line with the Warwicks in support, and here the Brigadier was ordered to be across the Oise at Pont l'Evêque by 6 on the morning of the 30th, the bridges being blown up behind him.

Leaving the vicinity of the river the Brigade marched by Carlepont, Champ-du-Merlier, Tracy le Mont, Bascule and Berneuil-sur-Aisne to Genoncourt, where the troops bivouacked, having been on the move for about fifteen hours, and this being the seventh night with no sleep beyond a few hours snatched here and there.

On the morning of August 31st the cheering news was received of the French success at the battle of Guise, and the march next day was a pleasanter one, the dust-laden roads of the preceding days being exchanged for the grassy alleys of the Forest of Compiègne, and the Division moving in two parallel columns, that on the right containing the 10th and 12th Brigades. The bivouac this night was beyond Verberie.

A very early start was made on the morning of September 1st, the 10th Brigade having received orders to proceed southwards to St. Vaast, but the march had scarcely commenced when news was obtained of the action at Néry, and first the newly-formed Composite Battalion of Warwickshires and Royal Dublin Fusiliers and then the two other units of the Brigade moved on the northern end of the village of Néry. By the time this was reached, however, the fighting was over, only a few rifle shots greeting the infantry as they neared the village, where General Haldane was directed to hold the ground and cover the collection and removal of the wounded. The 10th Brigade accordingly occupied a front from Néry to La Boissière Farm, withdrawing about 3 p.m. towards Rully *en route* for Baron, where the halt was to be made for the night.

This day Lieutenant-General Pulteney arrived from England to take command of the IIIrd Corps which was to include the 4th Division.

On approaching Baron there was some enemy rifle-fire from a large wood in the vicinity, but this annoyance very soon ceased, and the night passed quietly until 3 a.m. on September 2nd, when the march was resumed to Dammartin-en-Goële, situated on an isolated hill; this was reached before midday, but the troops were warned to march again at midnight.

As a matter of fact, however, the resumption of the rearward march was delayed, and it was not actually until 4.30 a.m. on the 3rd that the 10th Brigade, forming the rear-guard of the Division with a body of French cavalry, was able to move off. The march was quite unmolested and, tramping along by way of St. Mard, Juilly, Nantouillet, St. Mesmes, Messy, Claye, Souilly, Annet, Thorigny—near which the Marne was crossed—the troops reached Lagny, and the Battalion went into bivouac to the east of the Bois de Chigny, arriving there at 3.15 p.m.

On the evening of the 4th the Brigade moved some three miles only to a fresh bivouac at Magny-le-Hongre, which was left again in the early hours of the 5th, and the troops marched by Jossigny and Ferrières and through the Forest of Armainvilliers to Chevry where a halt was made and where the welcome news was received that, provided all went well with the French Fifth and Sixth Armies which were then engaged with the enemy, the retreat might be stayed and the advance be commenced. " We had marched without a day's halt for 13 days," writes Brigadier-General Haldane, " and had covered a distance of 170 miles. This, though not in actual mileage a feat, has some claim to be considered one when it is remembered that the majority of the troops were men who had returned from civil life and had not probably done much in the way of walking exercise or carrying weights for at least a year. The weather, too, was considerably hotter than is generally experienced in England, and what was perhaps the greatest trial of all was the lack of sleep. Indeed, except in the bivouac near the Bois de Chigny on September 3rd, no officer or man of the Brigade had had a full night's rest since August 20th, the day before the troops left Harrow for the front. The *morale* of all ranks, considering that they had been retiring for so many days, left nothing to be desired."

September 6th was a day to be remembered for all time, for the retreat had now come to an end, the Allied armies had not only turned about but were advancing against an enemy who in his turn was falling back ; " on that day " wrote Field-Marshal French,* " it may be said that a great battle opened on a front extending from Ermenonville, which was just in front of the left flank of the Sixth French Army, through Lizy on the Marne, Maupertuis, which

*Despatch of September 7th, 1914.

was about the British centre, Courtecon, which was the left of the Fifth French Army, to Esternay and Charleville, the left of the Ninth Army under General Foch, and so along the front of the Ninth, Fourth and Third French Armies to a point north of the fortress of Verdun."

At Chevry all the battalions of the 10th Brigade had received reinforcements, many officers and men who had become separated from their units in the confused fighting of August 26th at Haucourt had rejoined, and the Warwickshires and Dublin Fusiliers had now ceased to form a composite battalion, and all ranks stepped out manfully, chafing at the apparent slowness of the advance, and only longing to come up with and give battle to the enemy. During the 6th, 7th and 8th the Division pressed forward hot on the tracks of the Germans, and on the last-named date had reached the high ground south of La Ferté, where the immediate situation seemed to be that the enemy was still south of the Marne on the right, and north of it on the left; while the 12th Brigade had secured a bridge over the Morin River at Courcelle, with the 11th Brigade on its left trying to reach the crossing over the Marne at La Ferté, where the houses on the further bank were held by determined enemy rear-guards whose machine guns commanded the bridge at short range.

On this day Sir John French saw Major-General Snow, commanding 4th Division, and told him that he had hoped to be able to see the brigade commanders and commanding officers of units to tell them that " the 4th Division had by its conduct and bearing saved the situation at the beginning of the retreat and that its work would rank as one of the finest feats of the British Army "—in allusion no doubt to the stand made at Le Cateau on August 26th.

The troops remained halted near La Ferté until late on the afternoon of the 9th when a short march was made to the little village of Grand Mont Menard where the headquarters of the 10th Brigade was established. Here it was learned that General Snow had met with a somewhat serious accident, his horse having fallen with and rolled over him. He eventually had to be invalided home, his place at the head of the 4th Division being taken first, temporarily, by Major-General Sir Henry Rawlinson, and later by Major-General Wilson ; the following messages were received by the units of the Division from their commander, the first dated La Carrière l'Evêque, September 23rd, 1914 :—

" On temporarily relinquishing the command of the 4th Division to Major-General Sir Henry Rawlinson, Major-General Snow wishes to express his great appreciation of the gallantry and endurance of all ranks of the Division since their arrival in France during the time which he has had the honour to command them.

" He also wishes to express his thanks to all officers and men for their loyalty and support both in peace and war since he took over command of the 4th Division."

The second is dated from hospital at Woolwich, November 4th. 1914 :—

" Major-General T. D'O. Snow in relinquishing permanently the command of the Division wishes to record his gratification in hearing on all sides of the gallant deeds of the 4th Division.

" In bidding the Division ' good-bye ' he congratulates it on obtaining for its commander a 4th Division man, Major-General Wilson, and one who has commanded it with such success in all the hard fighting it has experienced.

" Major-General Snow wishes all ranks the best of fortune."

On September 10th the Brigade crossed the Marne near Le Saussoy, and, marching up a steep hill to the Montreuil road, now found itself on the high rolling ground between the Marne and the Ourcq rivers ; passing through Chambardy and Dhuisy the Division halted at Coulombs, the advanced guard occupying Cerfroid and Vaux. It was reported that the Germans had been in occupation of these places on the night previous.

On September 11th the 10th Infantry Brigade was at Villers-le-Petit ; on the 12th, moving as advanced guard of the Division, it reached Septmonts, from the ridge above which the Germans holding trenches on the further side of the Aisne were shelled at a range of 9,000 yards by our heavy artillery ; on the next day, the 13th, the greater part of the 4th Division was over the Aisne, the 10th Brigade crossing at Venizel just before midnight and reaching Bucy-le-Long about 1 a.m. on the 14th. The 2nd Battalion Seaforth Highlanders and the 2nd Battalion Royal Dublin Fusiliers were now sent forward to fill up gaps in the front of the 11th Brigade which had crossed the Aisne some hours earlier and which was holding the high ground above Bucy-le-Long. On this day Second-Lieutenant B. McGuire was killed in action.

It was now tolerably apparent that the German retreat had come to an end and that the enemy was holding positions which had been previously selected and prepared for defence ; on the morning of the 15th, too, instructions were received that offensive action on the part of the British was for the present to cease and that the positions the troops were holding were to be strengthened.

The trench warfare, which was for so long to endure, may now be said to have commenced.

On the 16th the following Special Order was issued by the G.O.C. IIIrd Corps :—

" The Lieutenant-General Commanding wishes to-night to compliment the troops of the 4th Division on the initiative displayed by them in seizing their present position on the right bank of the Aisne on the night of September 12th, and on the courage and tenacity with which they have held it for two days and three nights in circumstances calculated to test to the utmost their steadfastness and endurance.

" He deeply regrets the sacrifice of valuable lives that has resulted, but he wishes to impress on the troops that this sacrifice has not been unavailing and that the part played by the Division in holding its ground has been of vital importance to the success of the operations as a whole.

" The situation to-night is such as to necessitate a further sustained effort to hold on as hitherto, and the Lieutenant-General Commanding has every faith in the determination and capacity of the troops to do so."

During the remainder of the month of September the Battalion remained in the neighbourhood of La Montagne Farm, taking its turn of duty in the trenches. The weather, which during the previous month had been so fine and warm, had now changed for the worst and there was incessant rain. The enemy's artillery was very active and there was constant expectation of attack; German snipers also had commenced to be busy and during the month the Battalion had one man killed and four wounded. Reinforcements were now, however, coming out from home at tolerably frequent intervals; thus, on the 20th Second-Lieutenant Kennedy arrived with 85 non-commissioned officers and men, on the 23rd 163 men of Section " D " Army Reserve joined under Second-Lieutenants Tarleton and Maffett, while on September 29th Second-Lieutenant Hallowes brought a reinforcement of 90 men, 47 of whom were men of the Battalion who had been cut off on August 26th and had managed to escape to the Channel ports and thence to England, the remaining 43 being special reservists.

Early in October Sir John French represented to Marshal Joffre that it would be well that the British Army should be withdrawn from the Aisne and take the position originally assigned to it on the left of the French forces, a move which, while materially shortening its line of communications, would at the same time give it the task of defending the Channel ports. Marshal Joffre agreed to the proposal, and the necessary steps were then taken to put it into force. The transfer actually began on October 3rd, the two cavalry divisions, moving first, proceeding by road, and these were followed by the IInd, IIIrd and Ist Corps in the order named, the infantry doing the journey by march, by train and in motor-buses, and the first corps to move was ready for action on the Aire-Bethune line on October 11th. " From the chalky uplands and the wooded slopes there was a sudden change to immense plains of clay, with slow, meandering, ditch-like streams, and all the hideous features of a great coal-field added to the drab monotony of Nature."

On October 5th orders came for the 10th Brigade to move; next day there arrived the French troops who were to take over the line; and at 2 a.m. on the 7th the relief was completed and the last companies of the Brigade left the ground which for three weeks they

THE RETREAT AND THE ADVANCE.

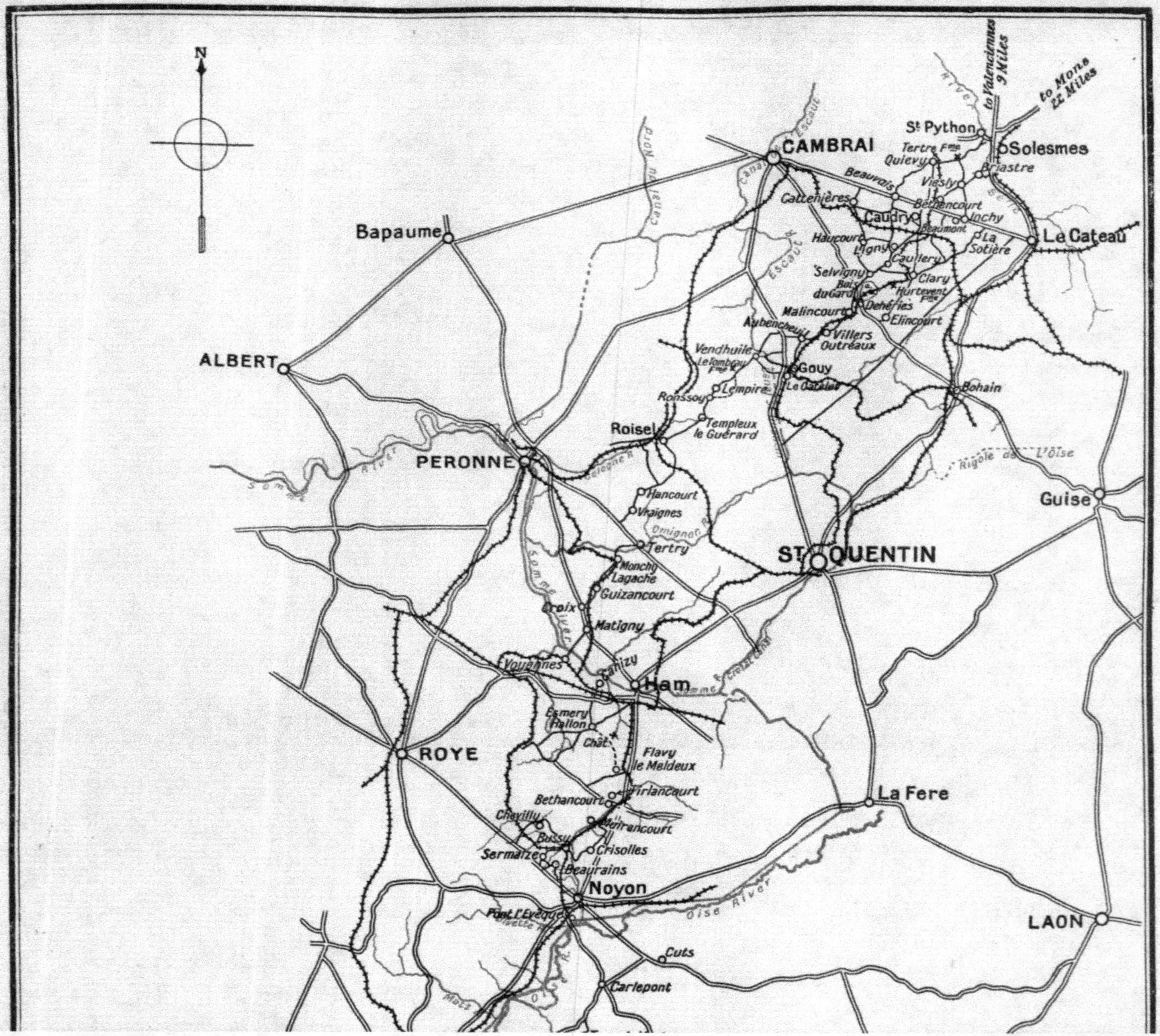

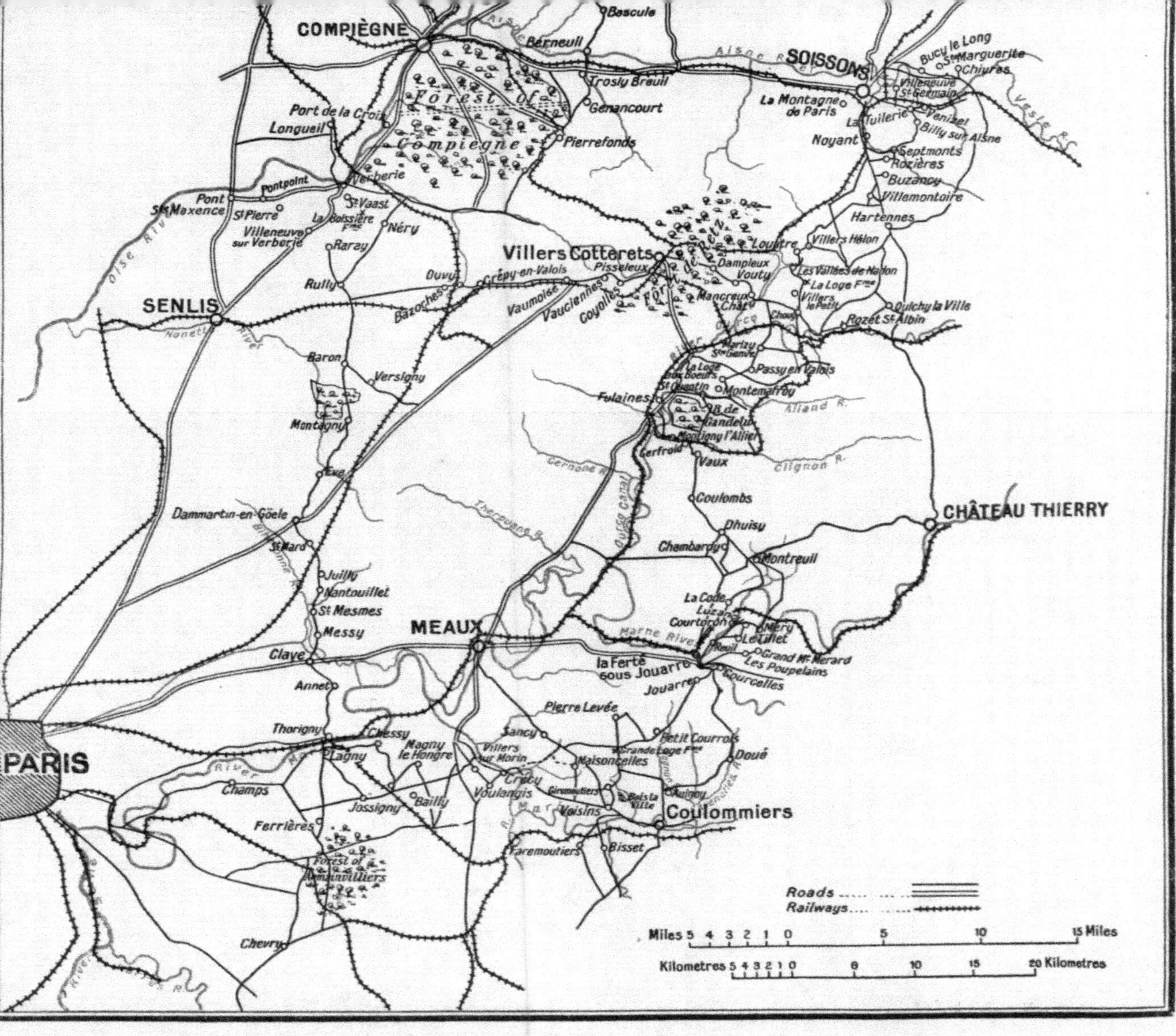

Reproduced from "A Brigade of the Old Army" (Lt.-Gen. Sir A. Haldane) by permission of the Publishers, Messrs. Edward Arnold & Co.)

had held on the right bank of the Aisne. The route followed was across the river at Venizel, then by Hartennes, Rozet St. Albin, and Coyelles to Rully ; here orders were received that the Brigade was to entrain, and the Royal Dublin Fusiliers then marched to Pontpoint where trucks were provided, and on the morning of Sunday the 11th the Battalion reached Hesdigneul, 4 miles south-east of Boulogne. Later in the day the journey was continued to Wizernes near St. Omer. On the following afternoon a large fleet of motor-buses came upon the scene and in these the troops were conveyed to Caestre.

On the morning of October 13th, the Brigade being now assembled, the troops marched off along the Bailleul road, but on approaching Flêtre, where the road was found to have been barricaded, it was reported that the ground immediately to the west of Meteren, rather less than $3\frac{1}{2}$ miles distant, was entrenched and held by the enemy.

A general attack by the 4th and 6th Divisions was now arranged to commence at 3.15 p.m., at which hour three battalions of the 10th Brigade began advancing direct on Meteren north of the road, while the 2nd Battalion Royal Dublin Fusiliers was sent against the enemy's right near Fontaine Houck. The frontal attack was successful in causing the enemy to leave his trenches, but the flank attack miscarried owing to the Battalion being drawn off in a wrong direction by a message from our cavalry, and consequently the village of Meteren was not occupied until late that night. The operations then came to an end, so far as the 10th Brigade was concerned, and the affair at Meteren was, as General Haldane remarks, " the last involving anything approaching to manœuvre that we were to take part in for many a day."

On the 14th the Brigade occupied Meteren, and Bailleul on the following day, while early on the morning of the 17th the 10th Brigade, on the right of the 4th Division, was ordered to move by Erquinghem on the River Lys and secure the large manufacturing town of Armentières, where the advance guard, composed of the Royal Irish Fusiliers, met with considerable opposition and sustained some loss ; while Houpelines, a suburb of Armentières, where the greater part of the Brigade was billeted, was heavily shelled.

" The ground," writes Brigadier-General Haldane,* " which, as right brigade of the 4th Division, I now held stretched from the River Lys, close to the hamlet of Le Ruage, in a south-easterly direction, where north of the main railway line to Lille I came in touch with the 6th Division. The area looking east and southwards is flat and seamed with ditches, mostly full of water, some of which, unworthily, rise to the dignity of being styled *rivières*. The frequent roads, main and other, within the district, bordered as they are by poplar trees, interfere with observation, and only by mounting to the top of tall chimneys or other lofty structures can a view, and that indifferent, be obtained. Rather less than a mile in front of my left and close to the River Lys, which here marks the frontier between France and Belgium, was the village of Frélinghien, which with the bridge across that stream lay within the German lines."

The Brigade was now very much below strength and so weak for the 2½ miles of front allotted to it that the commander, having obtained possession of the village of Frélinghien, was ordered to desist from further efforts to gain more ground.

The following complimentary orders were issued at the close of these operations :—

" The Field-Marshal Commanding-in-Chief the British Forces in the Field has instructed the General Officer Commanding the IIIrd Army Corps to convey to all ranks of the 4th Division his recognition and appreciation of their excellent work in the field during the operations of October 13th and 14th about Meteren and Bailleul. The successful execution of those attacks has had a most favourable effect on the general situation.

(Signed) B. BURNETT-HITCHCOCK, *Captain.*
For Lieutenant-Colonel G. S. 4th Division.

" NIEPPE, 17/10/14."

---

* *A Brigade of the Old Army*, p. 135.

HYDE PARK CORNER, PLOEGSTEERT WOOD.

A CONVENT IN YPRES.

" The G.O.C. wishes to congratulate the 4th Division most heartily on the tactical skill and fine fighting spirit shown by all ranks in to-day's successful operations. The news all round is excellent, and as the Indian troops are expected to arrive in line within the next few days there should be every chance of a successful termination to the present situation. He feels sure that the 4th Division will continue to-morrow the good work they have done to-day.

(*Signed*) A. A. MONTGOMERY, *Lieutenant-Colonel
Gen. Staff, 4th Division.*

" NIEPPE, 20/10/14."

On October 26th General Haldane had been promoted Major-General, and a few days later he learnt that he had been appointed to command the 3rd Division, then engaged to the east of Ypres, its commander, Sir Hubert Hamilton, having been killed some ten days previously. But as General Haldane's successor in the command of the 10th Brigade was not yet nominated, he remained on at Houpelines for another fortnight. On November 17th the 10th Brigade was relieved by the 19th and transferred to a part of the front not far from Ploegsteert, north of the River Lys, and here on the next day Major-General Haldane left to take up his new command, that of the 10th Brigade being assumed by Brigadier-General C. P. A. Hull.

Just before leaving the neighbourhood of Armentières a draft of 100 non-commissioned officers and men, mostly from the 5th Battalion of the Regiment, arrived under Lieutenant Hall, 3rd Battalion, and about the same time Second-Lieutenant Haines joined for duty.

The weather had now turned very cold, there was sharp frost with heavy falls of snow ; on November 19th the Battalion relieved the Rifle Brigade in the St. Yves position, finding it to be a pronounced salient, the trenches somewhat scattered and not of a very good type ; the enemy snipers were active and the trenches were heavily shelled ; and during the two days that the Battalion was up in the line eleven men were killed and seventeen wounded. On relief by the Warwicks the Fusiliers came into reserve and marched to Nieppe where they were billeted. In the last few days of the month the Battalion was in the trenches again and these, owing to the thaw which had now set in, were in a very ruinous condition and full of water.

At this time His Majesty the King paid a visit to his army in the field, and on his passing through Nieppe the Battalion helped to line the streets.

The trench warfare proceeded this month much the same as last, the misery of the life in the front line, where the trenches are described as being in " an appalling state," being in some measure mitigated by the efforts now being made to provide comforts, baths, fresh clothing and amusement for the troops when out of the line.

In this month Captain Smithwick and Lieutenant Bush joined headquarters, Lieutenant-Colonel Loveband* went home on leave which had now been opened, and Captain Frankland took over temporary command.

During December the Battalion had suffered more than forty casualties, two officers—Lieutenants Bush and French—being wounded, the former accidentally, while of the " other ranks " 16 were killed, 35 were wounded, and one man was missing.

There was quiet in the front line during the last few days of the first year of the Great War, and the battalion diary under date of December 31st, 1914, contains two words only—" No Sniping " !

*He had joined the Battalion on October 25th.

Bissezeele
Wylder
Bambecque STA
Oost Cappel
Rousbrugge-Haringhe
Esquelbecq STA
Wormhoudt
Herzeele
Houtkerque
Proven
Ledringhem
Watou
St Jan ter Biesen
Arneke
Oudezeele
Winnezeele
St Laurent
le Temple
Zermezeele
Hardifort
Steenvoorde
Abeele
Oudezeele
Noordpeene
Oeltezeele
Wemaers Cappel
Zuytpeene
Bavinchove
Oxelaere
CASSEL
Terdeghem
Godewaersvelde
St Marie Cappel
St Sylvestre Cappel
Eecke
2° 25'
30'
35'
40'
50° 55'
50'

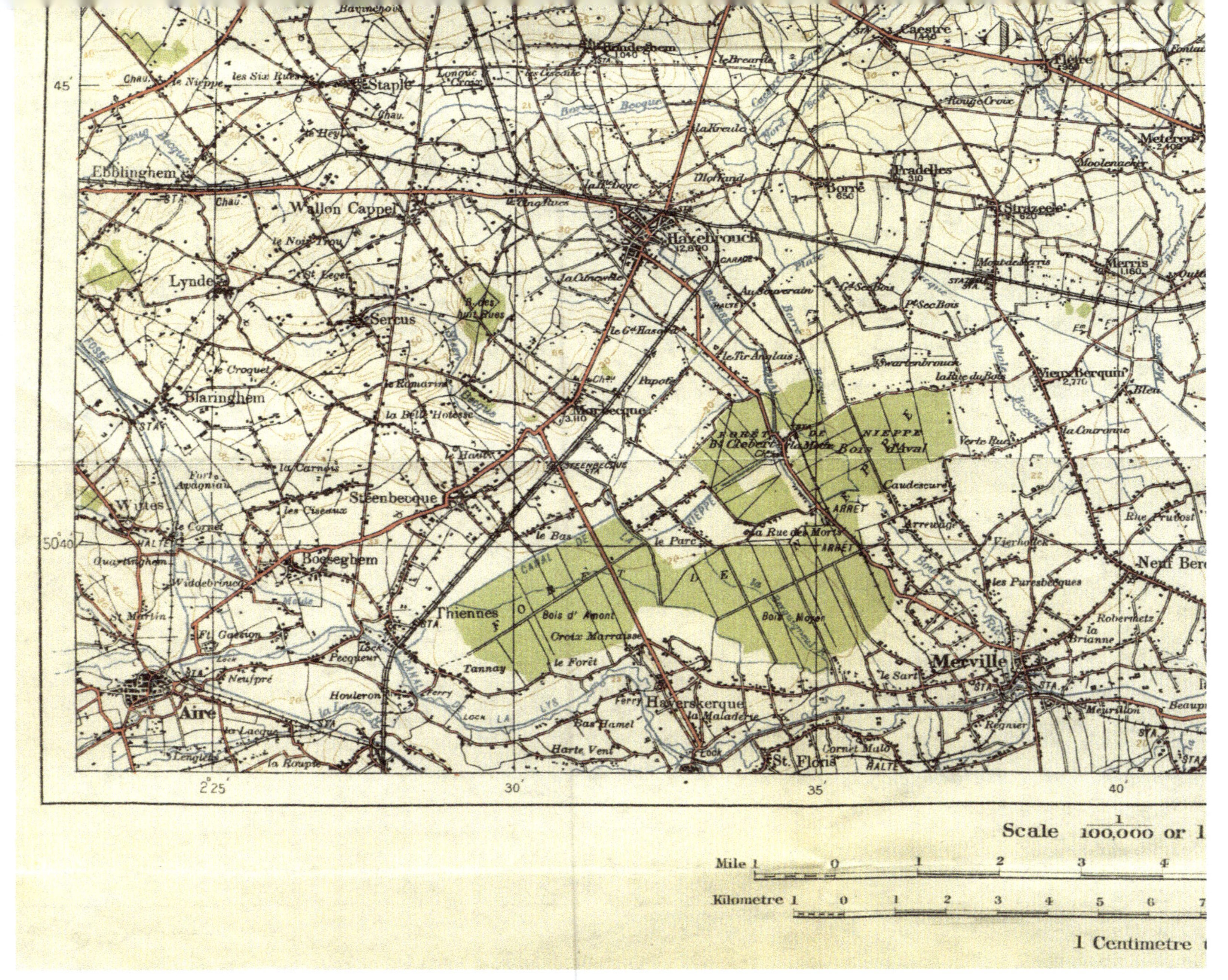

Bavinchove
Hondeghem
Caestre
le Brearde
Flêtre
Staple
Longue Croix
les Six Rues
le Nieppe
Chau.
le Hey
Rouge Croix
la Kreule
Metereu
Moolenacker
Ebblinghem
Ulofkand
Borre
Pradelles
Strazeele
Wallon Cappel
Hazebrouck
Merris
Lynde
St. Leger
B. des Sept Rues
Au Souverain
Gd Sec Bois
Pt Sec Bois
Sercus
le Gd Haton
Swartenbrouck
la Rue du Bois
Vieux Berquin
le Croquet
le Ramarin
Chau.
Papote
le Tir Anglais
Verte Rue
la Couronne
Blaringhem
la Belle Hôtesse
Morbecque
FORÊT DE NIEPPE
Bt Chebert la Motte Bois d'Aval
Caudescure
la Carnoÿ
Steenbecque
le Haut
ARRET
Arrewage
Vierhouck
Rue Probost
Wittes
les Ciseaux
le Bas
le Parc
la Rue des Morts
ARRET
Neuf Berquin
Quartinghem
Boeseghem
Widdebrouck
Thiennes
Bois d'Amont
Bois Moyen
les Puresbecques
St. Martin
Ft Gaëtion
Croix Marraisse
Robermetz
la Brianne
Aire
Neufpré
Houleron
Tannay
le Forêt
Merville
le Sart
St. Lengleti
la Roupie
Harte Vent
Haverskerque
la Maladerie
Meurillon
Beaup
Bas Hamel
St. Floris
Cornet Malo
Regnier

Scale 1/100,000 or 1
Mile 1 0 1 2 3 4
Kilometre 1 0 1 2 3 4 5 6 7
1 Centimetre

SALIENT
Poelcappelle
Langemarck
Zuydschoote
Steenstraate
Boesinghe
Pilkem
St. Julien
Zonnebeke
Elverdinghe
Woesten
Brielen
St. Jean
Wieltje
Frezenberg
Vlamertinghe
Kruisstraat
YPRES
Hooge
Poperinghe
Zillebeke
Gheluvelt
Veldhoek
Dickebusch
Voormezeele
Kruisstraat hoek
Kleinzillebeke
Reninghelst
St. Eloi
Zandvoorde
Westoutre
Hollebeke
La Clytte
Groote Verstraat
Kortewilde
Boescheppe
Wytschaete
Houthem
Berthen
Kemmel
Oosttaverne
Loere
Lindenhoek
45' 50' 55' 3°0'

Inch to 1·58 Miles
Reproduced by permission of the War Office, March, 1923.
10 Miles
15 Kilometres
to 1 Kilometre
Bailleul
Neuve Eglise
Wulverghem
Warneton
Bas Warneton
Warneton Sud et bas
Deulemont
Bois de Ploegsteer
Ploegsteert
le Oramp
les Ecluses
Prelinghien
Quesnoy sur Deul
la Crèche
Steenwerck
Nieppe
Houplines
Verlinghem
le Fulot
Armentières
Erquinghem
le Chapelle d'Armentières
Perenchies
Dompret
Sailly sur la Lys
Bac St Maur
Gris-Pot
Wez Macquart
la Bleue
Estaires
Fleurbaix
Bois-Grenier
Croix Marechal
Prémesques
Capinghem
Lomme
Ennetières en Weppes
Lavente
Escobecques
Englos

CHAPTER III

1915.

THE YPRES SALIENT.

DURING the greater part of the first three months of the year 1915 the Battalion remained in the Nieppe area, moving up to the front line and back again into reserve, and one reads daily in the war diary of heavy shelling, of intermittent sniping, of reports that the enemy has been heard engaged in mining ; but of major operations of any kind there is no mention.

" The winter fighting was commonly described as a war of attrition, but the phrase was a contradiction in terms. It was more correctly a period of waiting, a marking of time till further reserves in men and material were ready. But there was a positive side also to the Allies' plan. By frequent local attacks they kept the edge of their temper keen ; they prevented the enemy from concentrating in force against any part of their line ; they detained troops which might otherwise have been sent to Hindenburg. Their purpose was to be ready for any German attack, but to prevent it, if possible, by constantly worrying portions of the German front. The five hundred miles of the Allied line were held as to one-tenth by the British, and for the rest by the Belgians and French. It ran from Nieuport generally west of the Yser, along the Ypres Canal, in a salient in front of Ypres, behind Messines to just east of Armentières ; then west of Neuve Chapelle to Givenchy, across the La Bassée Canal, east of Vermelles, west of Lens, to just east of Arras. From Arras it lay by Albert and Noyon to Soissons, east along the Aisne to just north of Rheims, from Rheims by Vienne to Varennes, thence, making a wide curve round Verdun, to the west bank of the Meuse opposite St. Mihiel, and so on to Pont-à-Mousson on the Moselle. Thence it passed east of Lunéville to just east of St. Dié, ten miles inside the frontier. It reached the crest of the Vosges about the Col du Bonhomme, and then ran in German territory to Belfort and the Swiss border. . . . The winter's record was a chronicle of small things—a sandhill won east of Nieuport, a trench or two at Ypres, a corner of a brickfield at La Bassée, a few hundred yards near Arras, a farm on the Oise, a mile in northern Champagne, a coppice in the Argonne, a hillock on the Meuse, part of a wood on the Moselle, some of the high glens in the Vosges, and a village or two in Alsace. But these minute advances had their moral value for the troops engaged, and even a certain strategical importance for the campaign."*

Towards the middle of March rumours were heard that the 10th Brigade on relief by the 84th would move to an " unknown destination " ; later, that the Battalion was to move to Houplines and take over a portion of the line there held by the 18th Brigade. But these reports came to nothing, until on the night of March 20th, when everybody had settled down again to the old routine, the Royal Dublin Fusiliers received orders to take up a line north of the River Douve in conjunction with the Warwickshire Regiment, and at the end of the month the move was carried out, the new line occupied running from the River Douve to the Wulverghem— Messines road inclusive. The trenches here were constructed on the breastwork principle : the parapets were rather low, while, since the enemy lines were above those of the British, every movement by day was at once visible to the Germans. The reserve billets were at La Hutte.

The stay here was only a very brief one, for on the evening of April 12th the Battalion, having been relieved by a battalion of the 143rd Infantry Brigade (48th Division), marched to billets at Bailleul, over which town a Zeppelin passed during the night dropping a dozen bombs, failing to harm any combatants, but killing 3 women, 1 child, and 7 horses. Here on the 17th the Battalion was inspected by General Sir W. Pulteney, commanding IIIrd Army Corps, who said that he " wished to thank all ranks for the way in which they had held on to their trenches in the

* Buchan, *A History of the Great War*, Vol. I, pp. 457, 458.

1.—Chateau la Hutte, Hill 63, 1914-15.
2.—St. Ives, 1914-15.
3.—Battalion H.Q., "The Boreen," St. Ives, 1914-15.
4.—Water Tower, Ploegsteert, 1914-15.
5.—Major S. G. de C. Wheeler and Captain R. M. Watson, D.S.O., St. Ives, 1914.

vicinity of Ploegsteert Wood, that no one knew better than he did how arduous their duties were, and he congratulated them on the appearance and physique of the Battalion."

On April 23rd the 2nd Battalion Royal Dublin Fusiliers was ordered to be ready to move at half an hour's notice, and accordingly it left Bailleul at 7.30 that evening, marching north, billeted that night at Westoutre, and, leaving early again on the 24th, moved by Hensken, Zevecoten, Ouderdom, and Vlamertinghe to the outskirts of Ypres, which were reached at 8 p.m. At Ypres packs were discarded, and at midnight the Battalion marched on St. Jean, deploying at 4 a.m. on the 25th west of the Wieltje—St. Julien road.

The need that had arisen to call for the 10th Brigade to take hurriedly its part in what is known as the Second Battle of Ypres, was due to the success which had been won by the Germans in their attack on the 22nd upon the position held by the Canadians and the French between Bixschoote and Langemarck, owing to their unexpected employment of gas, resulting in the forcing of a gap five miles wide in the front of the Allied position.

Every hour of the 24th was an hour of danger where, while reinforcements were being rushed up, fresh ground had perforce been abandoned and heavy losses incurred.

" Reinforcements were now assembling to the immediate south of St. Julien. By evening the Northumberland Brigade and the Durham Light Infantry Brigade—both of the 50th Territorial Division—had reached Potijze. More experienced, but not more eager, was Hull's Regular Brigade, which had come swiftly from the Armentières region. All these troops, together with Geddes' detachment and two battalions of the York and Durham Territorials, were placed under the hand of General Alderson for the purpose of a strong counter-attack upon St. Julien . . .

" The advance was made at 6.30 in the morning of April 25th, General Hull being in immediate control of the attack. It was made in the first instance by the 10th Brigade and the 1st Royal Irish from the 82nd Brigade. . . . Little progress was made, however, and it became clear that there was not weight enough behind the advance to crush a way through the obstacles in front."*

"It was a desperately difficult undertaking," writes another historian.† "The night was extremely dark, the ground, which had not been reconnoitred, was honeycombed with trenches and strewn with barbed wire, and, moreover, the artillery had not been able to ' register '—that is to say, get its range of the *terrain*. Just before the attack was launched word came back that some Canadians were still holding out in the village of St. Julien," whence a fresh gas attack had driven the Canadian left on the morning of the 24th. "Therefore the place could not be shelled. The guns, however, opened on the wood west of the village. . . .

"The 7th Argyll and Sutherland Highlanders, a Territorial battalion that was on its trial that day, led with splendid dash on the right, the 1st Warwicks on the left. They were followed on right and left respectively by the 1st Royal Irish Fusiliers and the 2nd Dublins, while the 2nd Seaforths were ordered to connect with General Riddell's brigade of the Northumberland Division, which had been sent up to relieve the Canadians.

"As soon as our men got out of their trenches they were met by a terrific machine-gun and rifle fire at close quarters, whilst the German heavy guns in the rear spouted a continual torrent of shells over the fields through which the assault was delivered. Our men dropped left and right, but they never wavered, and the Irish Fusiliers and the Dublins, Irishmen all, fighting shoulder to shoulder, actually got into the outskirts of St. Julien." (In a letter from Captain Dickie of the 2nd Battalion he says :—" Captain Tobin Maunsell, with Corporal (now Sergeant) Lalor and a few men, got into the village up a ditch, but the rest of the company under me was mopped up trying to get there in extended order. Young French and Salvesen were killed next me and Elsworthy wounded, close to the village. I was hit, but managed to carry on for half an hour and arrange the defence of the last farm on the road up to the village.")

* Conan Doyle, Vol. II, pp. 62, 63.
† Williams, *With our Army in Flanders*, p. 73 *et seq.*

" The scattered ruins, the maze of trenches, and the barbed wire strung out everywhere, seriously delayed these two battalions and checked our advance. Two battalions of a brigade of the Northumberland Division, supporting the Dublins, lost their direction. . . . On the left the Warwicks and on the other flank the Highlanders got to within 70 yards of the German trenches in front of the wood. Here they were hung up and were ' properly hammered,' in the words of one who was there, by German high-explosive shells. Nevertheless, by this gallant attack the gap between the Canadians east of St. Julien and north of Fortuin was filled."

Of Colonel Loveband earlier in the action an officer writes as follows :—" One unforgettable scene remains in the writer's memory ; one company, which had lost all its officers and which had been ordered to retire, was doing so in disorder when the small, untidy figure of Colonel Loveband, clad in an ancient ' British warm ' and carrying a blackthorn stick, approached quietly across the open, making as he walked the lie-down signal with the stick. The effect was instantaneous, and for hundreds of yards along the front the men dropped and used their entrenching tools."

At the close of the first day's fighting Lieutenant-Colonel Loveband was wounded, and handed over command of the Battalion to Captain Bankes, and the Royal Dublin Fusiliers dug themselves in on a line facing, and a quarter of a mile from, St. Julien ; but all through April 26th to 30th the enemy kept up a tremendous bombardment right round the curve of the salient, making full use of their artillery superiority, and sweeping the trenches and all the roads through Ypres with a never-ending deluge of heavy projectiles.

In the meantime the French had been counter-attacking with a view of easing the pressure on their allies, but had made no progress, and Sir John French now decided that he could no longer afford to hold to our present exposed position ; he therefore ordered General Plumer, who was charged with the defence of the salient, to fall back upon a line prepared in the rear. Before, however, the withdrawal had commenced, the Germans, on May 2nd, launched a gas attack from St. Julien against the 10th, 11th, and 12th

Brigades holding the line in front of St. Julien and down to Fortuin ; but by this time the British were supplied with gas-respirators—of a sort—and the German attack failed to reach our trenches.   Fighting and intermittent shelling went on all through May 2nd and 3rd, and on the night of this last day the retirement began ; but it was not until the 4th that the Battalion withdrew from the line and, passing through the new one established in rear, bivouacked on the east bank of the Canal half a mile north-west of La Brique, arriving there about 2 a.m. on the 5th, and being greeted by two bombs dropped near the canal bridge from a hostile aeroplane, whereby two men were slightly wounded.

One of the surviving officers of this day's fighting writes : " One of the many instances of unrecorded gallantry and determination to put ' the Regiment ' first at all costs came to my notice in this action.   Lieutenant J. R. F. Hall, a bright, energetic and fearless young lad, who had survived the tedious though strenuous winter existence in the trenches, was sent back from the front line with a party to bring up rations from near St. Jean, just outside Ypres.   When he reached the dump it was easy to see that he was greatly in need of even only a few hours' rest, and I got leave for him to remain for the night at the advanced Brigade Headquarters, and for somebody else to take the party back to the line.   But I had done this without consulting Hall, who was furious when he heard of my action, and although almost exhausted from want of sleep and general overwork, pulled himself together and begged permission to be allowed to return with his party, using every possible argument, and finally gaining his point by saying—' Now, what's the use of being in a good regiment if you can't stick it out ? ' He accordingly returned that night to the line in command of his men, and was killed during a gas attack in a later stage of the same action."

In the operations which on this date came to a temporary conclusion the 10th Infantry Brigade had suffered more casualties than any other unit engaged, the battalions composing it having lost no fewer than 63 officers and 2,300 men, a very high proportion of

1.—Potijze, 1915.
2.—Dressing Station, St. Jean, Second Battle of Ypres, 1915.
3.—Lunch in Grounds, Vlamertinghe Chateau, when refitting, Second Battle
of Ypres, 1915.
4.—St. Jean, 1915.
5.—Temporary "Dug Out," Canal Bank, Ypres, 1915.

their total numbers. The 2nd Battalion Royal Dublin Fusiliers had 7 officers killed, 9 officers wounded. The names of the officers who were killed are as follows :—Captains E. N. Bankes and F. N. Le Mesurier ; Lieutenant C. S. French ; Second-Lieutenants F. Sparrow, C. W. Peel, E. M. Salvesen and W. White ; the wounded officers were Lieutenant-Colonel A. Loveband, C.M.G. ; Captains S. G. de C. Wheeler, R. M. Watson, D.S.O., and R. J. D. Johnson ; Lieutenants J. MacN. Dickie, A. L. Elsworthy and F. Treacher ; Second-Lieutenants E. F. Radcliffe and H. L. Ridley.

It will be remembered that while the 2nd Battalion was engaged and suffered these heavy casualties in Flanders, the 1st Battalion, in the landing at Cape Helles in the Dardanelles was equally losing many of its best and bravest.

On May 5th the 2nd Battalion Royal Dublin Fusiliers was moved at dusk to a bivouac round the Chateau des Trois Tours, and the two following days were passed in comparative quiet. But at noon on May 8th the Battalion was warned to be prepared to support the 84th Brigade of the 28th Division, which was in front of Frezenberg, where the Germans, assisted by the employment of gas and the extreme weight of their advance, had driven a gap in the British line.

At 1.30 p.m. on Saturday, the 8th, the Battalion left its bivouac, joined the 1st Battalion Warwicks outside Ypres, and marched in artillery formation through La Brique and St. Jean to the woods near Potijze Chateau, coming under heavy shell fire. These two battalions joined the 1st York and Lancaster, the 3rd Middlesex, and the 2nd East Surrey in a strong counter-attack which reached Frezenberg, but was eventually driven back, and finally remained on a line running north and south through Verlorenhoek, the Battalion on the early morning of the 9th occupying a line with the left 150 yards south-east of Shell Trap Farm, with one company in reserve in trenches north of Potijze Wood. The shelling by the German guns that day was the worst that the Battalion had so far experienced in the war, while this, with the accuracy of the enemy sniping and the closeness of the machine-gun fire, made the reconstruction of trenches that night a work of great difficulty.

The Battalion held its ground until the night of the 12th, when it was relieved by the 15th Hussars and the London Rifle Brigade, and marched back through St. Jean and Ypres to bivouac in the grounds of Vlamertinghe Chateau.

" May 13th may be reckoned the last day of the second battle of Ypres. . . . It was not a battle like the first battle of Ypres, when our men met the flower of the Prussian Army face to face, and withstood a succession of onslaughts delivered with an incredible disregard of human life. The second battle of Ypres was a battle of machinery, in which the German infantry skulked behind their gas-cylinders and machine guns, and waited for their heavy guns to prepare for them victory at a cheap price."*

For ten days only, however, while the British repaired their battered lines, was there a lull in the fighting, but there was no change in the German plan of campaign, and the fighting which had ceased on the 13th broke out anew on the 24th, on which day the German poison gas came " drifting down wind in a solid bank some three miles in length and forty feet in depth, bleaching the grass, blighting the trees, and leaving a broad scar of destruction behind it " ; while a tornado of shells fell on the British trenches. The main force of the gas cloud struck the extreme right of the 4th Division, where was the 10th Infantry Brigade, the 12th being on the left of the line, and the 11th Brigade being in reserve. On the 10th and 12th, therefore, fell the full weight of the attack.

The following report on all that this day occurred, so far as the Battalion is concerned, was drawn up by Captain Leahy, the only officer left with the 2nd Battalion Royal Dublin Fusiliers when it was finally taken out of action.

" Colonel Loveband, Major Magan, second in command, Russell,† R.A.M.C., and I, acting adjutant, had just finished dinner

---

* Williams, pp. 95, 96.

† This officer had, on September 10th, 1914, taken the place of Captain Davey, R.A.M.C., taken prisoner on August 26th.

in our headquarter dug-out at 2.30 a.m. Previous to this the
Colonel and Magan had been round all the front line trenches and
spent considerable time in Shell Trap Farm. Something suggested
' gas ' to the Colonel during his round of the trenches as he personally
warned all company officers to be prepared, and Russell had inspected all the Vermoral sprayers and warned each company about
damping their respirators. There were ten sprayers in working
order that night—one with each machine gun and the remainder
distributed along the trenches.

" At about 2.45 a.m. the Colonel and I were standing outside
our dug-out, some 400 yards behind the first line of trenches, looking
in the direction of Shell Trap Farm, when we saw a red light thrown
up in the German lines to the north-west of the Farm, and immediately three lights (red) were seen directly over Shell Trap Farm and
a few more lights (red) in the German lines from the direction of
C.30—south-east from where we were standing. A few seconds
later a dull roar was heard—more like an explosion, certainly not a
shell—and we saw the gas coming on either side of Shell Trap
Farm. The Colonel shouted, ' Get your respirators, Boys, here
comes the gas ! ' We had only just time to get our respirators on
before the gas was over us—the doctor, Russell, who was seeing to
other people got some gas before his own respirator was adjusted.

" In the trenches the ' Stand-to ' was just over and rum was
being issued, so there could not have been any element of surprise
other than the sudden appearance of gas ; everyone was awake.
There was a very gentle breeze, the gas was very dense and took
considerable time to pass over—about three-quarters of an hour.
From the nature of the ground—a gradual slope towards Battalion
Headquarters from the first line of trenches—I do not think that the
gas lasted as long over the trenches as it did over us at Headquarters,
or the troops (two companies 9th Argylls) in the retrenchment.
Also our right company, ' A,' under Captain Basil Maclear, did not
get such a heavy gassing as the remainder. While the gas was at
its height with us some of the 9th Argylls retired from the retrenchment. Colonel Loveband and I ran out to stop them, there were

five Dublins with them who came to us and three of them were employed later as orderlies, the other two were gas-affected. . . . By this time the gas was clearing a little but was still fairly thick, and Russell drew our attention to the trench on the left of Shell Trap Farm, out of which men were pouring, and we almost immediately saw three Germans in the right corner of Shell Trap Farm, close against the buildings.

" The Colonel sent the following messages :—

" 1. To 9th Argylls' Retrenchment :—' Enemy are in Shell Trap Farm, counter-attack at once.' (4.45 a.m.)

" 2. A verbal message to King's Own on left of 18th Royal Irish :—' Enemy are occupying some of the 18th Royal Irish trenches and there are a few in Shell Trap Farm '—or orders to that effect, it was a verbal message.

" 3. To O.C. 9th Argylls :—' Enemy are in Shell Trap Farm Send forward two companies to reinforce retrenchment line and counter-attack at once.' (4.45 a.m.)

" Almost immediately we received a verbal reply from the King's Own saying they could deal with the enemy in 18th Royal Irish trench if we could deal with those in the Farm—or words to that effect. We had also asked the King's Own to inform the 10th Infantry Brigade as our wires by that time were all cut ; just before they were cut the King's Own sent another message saying they were all right.

" By this time the gas had quite cleared and the enemy were shelling heavily, a good many of the 9th Argylls in the retrenchment had moved up, but had gone rather in support of ' A ' and ' B ' Companies, viz., to our right and centre trenches instead of to Shell Trap Farm, and I think an Argyll officer with machine gun and detachment had moved up to one of our support trenches, but to which trench I cannot say. Verbal messages then started coming in from the front trenches—all full of confidence ; and Basil Maclear told us about the enemy in Shell Trap Farm. This was the first time he could see them there as they had previously kept under

1.—Group, before going into action, Second Battle of Ypres, 1915.
2.—Lt.-Col. A. Loveband, C.M.G., and Capt. Russell, R.A.M.C., in trenches, Houplines, 1914.
3.—Lt.-Col. A. Loveband, C.M.G., Major T. Magan, Capt. Carroll-Leahy, M.C., outside Battalion H.Q. (The Grouse Butt), Ypres, May 23rd, 1915.
4.—Reserve Trenches, St. Jean, April, 1915.

cover of the buildings out of his sight. The Colonel then sent a second message to O.C. 9th Argylls :—

" ' Enemy are in Shell Trap Farm, please send up two companies at once and counter-attack and reoccupy Farm. Germans are in right side of farm. Inform 10th Infantry Brigade.' (5.55 a.m.)

" Meanwhile *heavies* were being dumped into our trenches and there was a severe enfilade machine-gun fire opened on us from the Royal Irish trenches to the left of the farm. Getting messages down to Battalion Headquarters was a difficulty and it was impossible to get an orderly back to the divisional support line without his being badly hit—every orderly who came to and went from us was hit, yet every time there was a message to go there was a volunteer to take it. My orderlies were used up by then so I had to use gassed men and signallers.

" The Colonel then sent the following message to the 10th Infantry Brigade :—' The Germans are in Shell Trap Farm and I can see a few in small building on right of farm. Get artillery to shell it. I believe my two platoons are still holding trench to left of farm. The Germans have occupied the right trench of 18th Royal Irish. Have been unable to get up the two companies of Argylls in support line so far. Argylls in retrenchment have moved up, but I fear to my centre and right, and am not sure if any went up to the farm. Reinforcements are required. Situation not satisfactory.'

" This message was untimed but I think it was sent about a quarter of an hour before Burt-Marshall* arrived with us. The enemy were now using gas-shells which disturbed our eyes but did not seem to have any other effect. Meanwhile the Colonel, Russell and I were standing at the back of our dug-out, Major Magan was inside gassed and out of action since about 4 a.m., and Russell was hit by a piece of shell, but remained at duty and did most gallant work throughout.

---

* Of the Signal Company.

" Colonel Poole* then came up and Colonel Loveband explained the situation. As far as I can remember Colonel Poole had not gone very long before Burt-Marshall arrived, and he, the Colonel and I were standing at the back of the dug-out—the doctor was inside—when bullets came from behind and presently Colonel Loveband was hit through the heart ; he died without a word though he tried to say something. Marshall was hit in the shoulder from a bullet coming from the same direction ; he was most plucky, wouldn't wait, but raced off to stop the firing. Our guns were shelling Shell Trap Farm heavily by this and our shrapnel appeared to be bursting well clear of our trenches. The enemy recommenced their shelling and ' heavies ' were being dumped into our right trenches, while the rifle and machine-gun fire was very accurate from the 18th Royal Irish trenches which the Germans held in force. It was marvellous how quickly they converted the parapet and everything was done without confusion and with proper method.

" Maclear then sent the following message :—' Must have more men on left between " B " Company and Shell Trap Farm.' This particular spot was occupied by ' D ' Company and they were in a bad plight, for it was now obvious that the 9th Argyll and Sutherland Highlanders who had supported us had gone too much to their right, very few had supported ' D ' Company or ' C ' Company, with Major Johnson, in front of the farm.

" Some men now began retiring on the right and Russell and I ran out to stop them. Russell was perfectly splendid and nobody could have given more help, but it was useless. A few of our men were with those retiring but these stopped when they saw their own officers and obeyed orders ; the remainder were dropping like flies and very few reached the divisional support line. . . . On arriving back at the dug-out I found Lieutenant Tarleton had just managed to crawl down with a message from ' D ' Company—Captain Barry, who had sent Tarleton down with the message as he was badly wounded in the trenches ; ' D ' Company was in a bad way with

* Royal Warwickshire Regiment.

only one man per five yards of front capable of using a rifle and this with all the supports in the front line.

" Tarleton was mad with gas and badly hit, but he had the heart of a lion and delivered his message—it was the only thing on him which was not covered with blood and mud.   Later in the day when he had somewhat recovered there seemed to be something on his mind, but what it was we could not discover, he was too done to talk, but when we produced his Véry pistol from his haversack he was quite happy—he knew that Véry pistols were very precious !

" A message arrived now from Second-Lieutenant Wright, 9th Argyll and Sutherland Highlanders :—' I have a machine-gun and a few men 9th Argylls and Royal Irish Fusiliers in the support trench which the guide will show you—almost due west of the farm.   Are you going to relieve us or shall we retire ? '   I sent him a message, telling him to hold on and that I expected supports.   The Warwicks had a few men coming up on the left of where I was.   Wright evidently mistook the few 18th Royal Irish men he had with him for Royal Irish Fusiliers.

" Shanks, our machine-gun officer, sent the following message : —' Enemy are strongly entrenching themselves between Shell Trap Farm and ' D ' Company trenches.   We are extremely weak. Reinforcements are urgently needed.   Reference attached message, two companies did reinforce us but they have been considerably weakened.'

" Maclear sent the following :—' Very many of our men are surrounded.   We must have reinforcements.'   Maclear was on the left of ' A ' Company and right of ' B ' Company, and when the Germans got into ' D ' Company trench, as they did about this time, he gallantly led a party to bomb them out, and was killed doing this.

" The next message came from Second-Lieutenant Kempston, who was with ' B ' Company (Second-Lieutenant Hall had been killed), it ran :—' For God's sake send us some help, we are nearly done.'   Meanwhile I had scribbled other messages of which I have no copies and at 12.45 p.m. sent the following to the 10th Infantry Brigade :—' Reinforce or all is lost.'

" Shanks, our machine-gun officer, sent one as follows :—' Our artillery are firing into our trenches.   Please hasten reinforcements ; they are needed all along the line.   The Germans hold our trenches from Shell Trap Farm for about 200 yards towards us.   Maclear, Thomas, and Hall are killed.'

" Just after that Kempston was killed, but Shanks was still fighting with ' B ' Company, and that is the last I heard of him. ' A ' Company were also holding out, presumably under Second-Lieutenant Young.   This officer was got away by somebody in the evening, as he died of wounds in hospital on the 25th.   Of ' C ' Company in front of Shell Trap Farm I heard little, but they were still fighting at 12 noon, though they were having a bad time of it from machine guns in the 18th Royal Irish trenches.   Major Johnson died of gas early in the morning, and Second-Lieutenant Considine was killed about 8 a.m. ; Second-Lieutenant Moran was with me about 2 p.m. and off his head from gas.

" When the wounded were sent away after dark, there were no Dublins in front of Battalion Headquarters.   From about 2.30 p.m. there was no fighting in our trenches ; everyone held on to them to the last ; there was no surrender, no quarter given or accepted ; they all died fighting at their posts.

" At 9.30 p.m. I received the following message :—' Please withdraw your headquarters and all men in the retrenchment if any are still there, and report at Brigade headquarters west of the Canal.' "

In accordance with the above, the Battalion—*strength, 1 officer and 20 other ranks*, all that remained out of 17 officers and 651 non-commissioned officers and men who had stood to arms in the morning—crossed the canal and bivouacked on the west bank one mile and a quarter west of La Brique, moving next day into the grounds of Mertingue Chateau.

Of the Battalion, there were killed in action 12 officers : the officers were Lieutenant-Colonel A. Loveband, Major R. J. D. Johnson ; Captains B. Maclear and J. H. Barry ; Lieutenant Moran ; Second-Lieutenants R. J. Kempston, M. O'C. Cuffey,

M. C. N. R. Young, J. R. F. Hall, D. G. Thomas, C. D. Considine, and F. G. K. Judd ; four officers—Major A. T. Magan, Lieutenant G. W. B. Tarleton, Second-Lieutenant R. M. Patterson, and Captain Russell, R.A.M.C.—were wounded, while Second-Lieutenant W. J. Shanks was missing.

Fortunately a draft was waiting to reinforce the Battalion, but even then it contained only 4 officers, including the chaplain and the medical officer, and 190 non-commissioned officers and men.

On May 28th Lieutenant-General Sir W. Pulteney, commanding IIIrd Army Corps, made an informal inspection of all that was left of the 2nd Battalion Royal Dublin Fusiliers, and remarked " how sad it was to see so few remaining, emphasizing the fact that they should console themselves with the knowledge that those who had gone had done their job."

Between June 1st to 4th, 5 officers and 223 men arrived and were taken on the strength of the Battalion ; and on the 7th another reinforcement of 2 officers and 125 other ranks joined : so that when on the 8th Major H. W. Higginson, from Brigade Major, 143rd Infantry Brigade, took over command from Captain Leahy, the Battalion had again attained a tolerably fair strength. But fortunately this particular area was now fairly quiet except for occasional heavy shelling, and the 38 casualties sustained during the month of June were more than made up for by the drafts which had been sent out from home, and which amounted to 19 officers and 629 non-commissioned officers and men between June 1st and 30th.

Early on the morning of July 6th the British guns opened fire on the German trenches which the 11th Brigade was about to assault. The enemy's reply was fairly steady on the front of the Battalion, and also on the communication trenches and the area in rear of the fire-trenches. At night the Battalion removed lengths of wire in the front of the trenches and showed bayonets at intervals, in order to give the impression to the enemy that an attack was preparing. The Germans put a good many shells over, and in the afternoon massed for an attempt to retake the trenches the 11th

Brigade had captured from them, but the gathering was dispersed by the British and French artillery, and by 7 p.m. all was fairly quiet. The casualties in the Battalion were, however, by no means negligible : Second-Lieutenant Conroy and 4 men were killed, while Second-Lieutenant Tittle and 33 other ranks were wounded.

On the morning of the 7th it was found that the previous day's shelling had done a good deal of damage to the parapets, which were as far as possible repaired. About 2 p.m. it was reported that the Germans were again massing for a counter-attack on the trenches they had lost on the 6th, and half an hour later the attack commenced, but was unsuccessful. During the afternoon there was a good deal of heavy gun-fire, the enemy firing a large number of 5.9's, one of which, unfortunately, fell into one of " C " Company's dug-outs, killing 4 men and wounding 2. One poor fellow had a leg, foot, and arm blown off, and lived for some time afterwards, while another had a leg blown off. The casualties from midday on July 4th to noon on July 8th then numbered 1 officer and 9 men killed, 2 officers and 53 men wounded, and 1 officer missing. The missing officer was Second-Lieutenant MacKeag, attached, who on the night of the 5th-6th was in charge of a party cutting grass in front of the left trenches about 300 yards from the German line. MacKeag went out towards the front, and was never seen or heard of again, though search, necessarily at night only, was repeatedly made for him.*

On the 9th the Battalion came out of the line and marched to rest billets at Houtkerque, where on the 14th Lieutenant-General Sir H. Plumer, Commanding Second Army, inspected the 10th Infantry Brigade, and specially complimented all ranks on their doings on April 25th ; he said :—

" *Colonel Poole, officers, non-commissioned officers and men of the 10th Brigade,—*

" *I much regret that your Brigadier, General Hull, is not present to-day, and if I could have deferred my visit I should have done so. This is a visit and not an inspection, though I feel bound to say that*

---

* It was afterwards ascertained that this officer had been taken prisoner.

had it been so the appearance of the men and their general turn-out could not have been better. After your two and a half months of hardship, and in spite of your severe fighting, you are able to turn out as soldiers, march as soldiers, and handle your arms as soldiers. I should like you to know the part you have been playing in this campaign since you came to the Ypres Salient.

" When our line was broken, through no fault of the British troops, two tasks devolved upon us. One was to repel the enemy attacks ; the other was to make frequent counter-attacks, which to the troops already engaged seemed meaningless, and which were necessarily very costly. The result was that we were able to maintain our front and withdraw to a fresh position at the time when it suited us to do so. When the history of this war comes to be written, the part which you have taken in holding that salient will be an episode not the least creditable among the many heroic deeds of the British Army.

" You are now leaving the VIth Corps, and I, its Army Commander, bid you farewell. I am sorry to lose you. You are going to form the nucleus of a new corps in a new army. Your task will be to give to the new units the benefit of your experience. If you succeed, as I am sure you will, in bringing them up to your high standard, you will form a highly efficient corps, and the General who commands you will be a fortunate man.

" I wish you all the best of luck."

On the 20th the Brigade was inspected by the Commander-in-Chief, Field-Marshal Sir John French, and two days later left the VIth Corps and the Second Army for the VIIth Corps and the Third Army. On the 4th Division leaving his command, Lieutenant-General Sir John Keir issued the following order :—

" I am debarred by orders from issuing any complimentary farewell order to the 4th Division, but I would be glad if you would take an opportunity to express verbally to units of the 4th Division my regret that they are leaving the VIth Corps. The 4th and 6th Divisions, which up till quite recently composed the IIIrd Corps, are about to be separated. The relations between them have always been of the most cordial nature, and their parting will, I know, be the cause

of much sorrow. In wishing the 4th Division good-bye, I desire to thank all ranks for the good work they have done on the front they have recently vacated, and also for their fine achievement in gaining an important position which has added much to the stability of the line.

" I wish the Division as a whole, and each of its members, a happy future."

The Battalion left Houtkerque on the afternoon of July 22nd, and moved by march-route and train via Godewaersvelde, Freschwiller and Vauchelles-les-Auttie to Bertrancourt which was reached on the 25th, and whence on the day following the Battalion moved up to the front line and took over the trenches from a French corps, the 3rd Battalion 64th Regiment. " The country in this district," writes an officer of the Fusiliers in his diary, " is a pleasant change from the Ypres Salient. It is a fine open country with rolling fields of corn with large woods here and there. There is little sign of war except that the countryside is very deserted, only old men and women and children are to be seen in the villages. The main roads are good with none of the *pavé* one meets in Belgium and Northern France."

The sector of front here allotted to the Battalion was just opposite the village of Serre, then held by the Germans ; the 5th Battalion Warwickshire Regiment* was on the left, and this part of the line had been captured from the enemy by the French early in June, and contained many very deep dug-outs some 20 feet deep ; the front line was well planned but the parapets poor.

Hereabouts the Battalion remained all through the month of August, but on September 2nd took over trenches east of Auchonvillers, a village which had suffered greatly from shell-fire at various times, but there remained still a few habitable houses among shady orchards. On the whole this was a very pleasant sector, the trenches were good, the weather for some time remained fine, and the enemy was very quiet. The men took great pride in their trenches and carried out excellent work on them, paving the main communication trench with bricks brought from Auchonvillers.

---

* 143rd Infantry Brigade, 48th Division.

ARRAS -
2° 20'   25'   30'   35'
Sibiville   HALTE
Houvigneul
Magnicourt-sur-Canche
Givenchy-le-Noble
Manin
Bellavesnes
Noyelette
Gou
Montenesc
Sericourt
Houvin-Houvigneul
Sars-lez-Bois
Lignereuil
Noyelle-Vion
Honval
Denier
Avesnes-le-Comte
St Hilaire
Frevent
Pit Bouret-sur-Canche
Cánettemont
Berlencourt
Bois de Faye
B. de Robermont
Beaufort
Hauteville
Briq
Wanquetin
Rebreuve
Brouilly
Ioncourt
Blavincourt
Apogrenée
Liencourt   ARRET
les Vents
Lattre-St. Quentin
Etrée-Wamin
Pavillon
B. des Onze
Simencourt
Rebreuvette
Wamin
le Cauroy
Grand Rullecourt
Gy R.
Fosseux
Chau
Fm Colonne
la Couture
Oppy
Arbre
Beaudricourt
Barly
Gouy-en-Artois
Beaume les Log
Mon Leblond
la Haie Grard Fm
Ivergny
Sus-St. Léger
Sombrin
Monchiet
Mon Fre
Warluzel
Soncamp
Bavincourt
B. de la Bouillère
le Bac du Sud
le Souich
Tuilé
les Annelles Fm
Baillenlval
Mon Fre de la Fontaine
Arbre
Forêt de Lucheux
Saulty
Larbret
Bailleulmont
Brevillers
Coullemont
Couturelle
Neuvillette
Bouquemaison
Bois de Robermont
Humbercourt
Gombremelle
Laherlière
Lucheux
Min Fm
Fm de la Breffaye
Solereneaule
la Cauchie
la Closerie Fm
Chau
la Bazèque Fm
les Calimonts Fm
Bois du Parc
la Folie Fm
la Bellevue
HALTE
la Bazèque
Berles
Ransart
Hte Visée
Gromies
Luchuel
Bois de Warron
le Gros Tison
Humbercamp
Pommier
HALTE
Bout des Prés
Pommera
Wamincourt les Pas
Gaudiempré
Bienvillers-au-Bois
Milly
l'Espérance
Mondicourt
Hem
le Collège STA
le Marais sec
Beaurepaire
HALTE
Grenas
Grincourt les Pas
Beaucamp Ru
St. Amand
CITADELLE
DOULLENS
Pit Caumesnil
Halloy
Hurtebise Fm
Ras
Bois du Chatelet
Bois de St. Pierre
Hénu
Souastre
Fonquevillers
la Bray

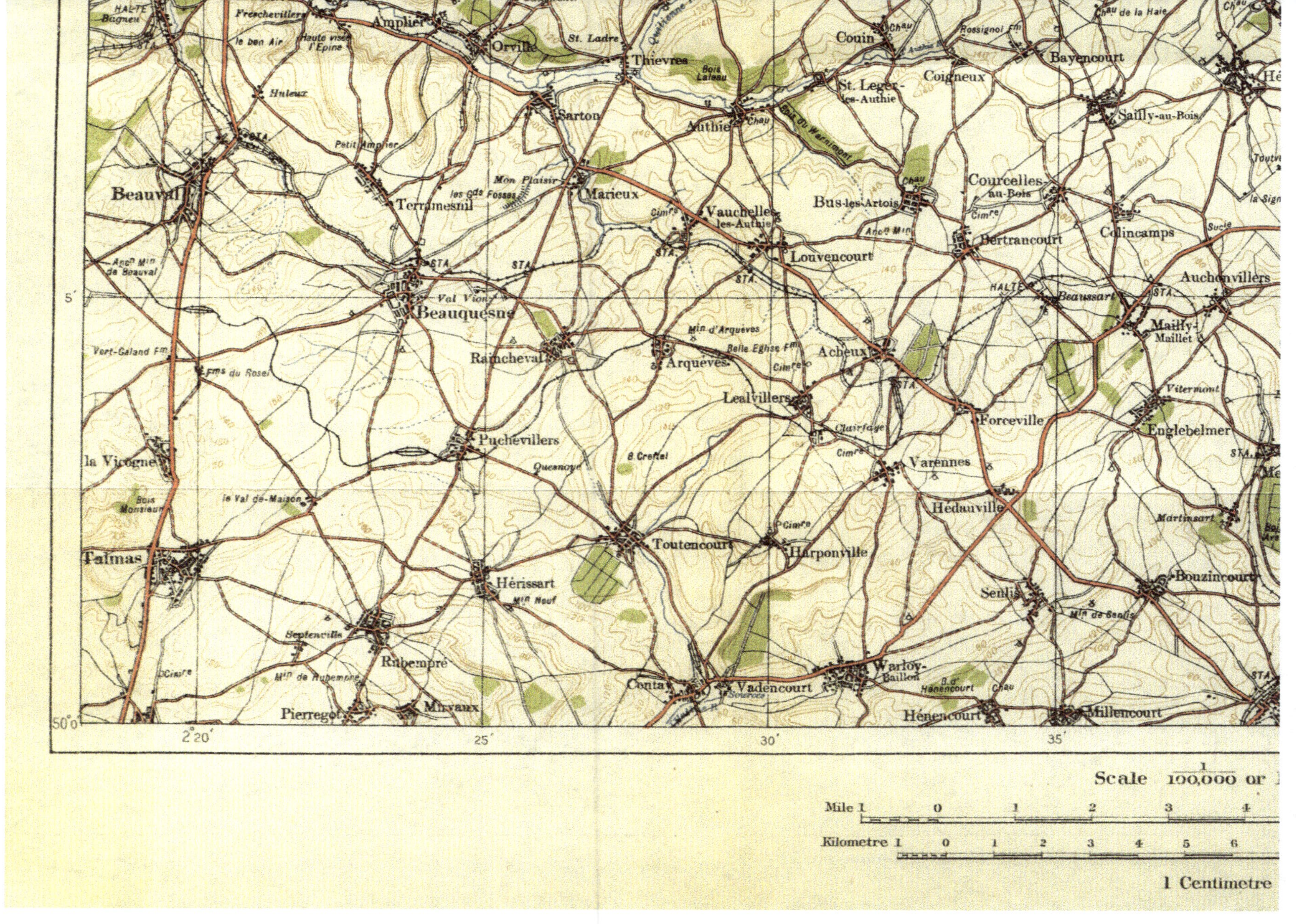

HALTE Bagneu
Freschevillers
le bon Air
Haute visée l'Epine
Amplier
Orville
St. Ladre
Thievres
Bois Laleau
Couin
Chau
Rossignol Fm
Bayencourt
Coigneux
St. Leger-les-Authie
Sailly-au-Bois
Huleux
Sarton
Authie
Chau
Bois du Warnimont
Courcelles-au-Bois
la Sign
Petit Amplier
Mon Plaisir
Marieux
Cimre
Vauchelles-les-Authie
Bus-les-Artois
Cimre
Sucie
Beauval
les gds Fosses
Terramesnil
Ancn Min
Bertrancourt
Colincamps
Ancn Min de Beauval
STA
STA
Louvencourt
STA
HALTE
Beaussart
STA
Auchonvillers
5'
Val Vion
Beauquesne
Min d'Arquéves
Mailly-Maillet
Vert-Galand Fm
Ramcheval
Arquéves
Belle Eglise Fm
Cimre
Acheux
STA
Vitermont
Fms du Rosel
Lealvillers
Clairfaye
Cimre
Forceville
Englebelmer
la Vicogne
Puchevillers
B Crestel
Varennes
STA
Bois Monsieur
le Val de-Maison
Queancye
Hedauville
Martinsart
Talmas
Toutencourt
P Cimre
Harponville
Cimre
Hérissart
Min Neuf
Senlis
Bouzincourt
Min de Senlis
Septenville
Warloy-Baillon
Rubempré
Min de Rubempré
Conta
Vadencourt
Source
B.d Hénencourt
Chau
STA
50°0'
Cimre
Pierregot
Mirvaux
Hénencourt
Millencourt
2°20'
25'
30'
35'
Scale 1/100,000 or 1
Mile 1   0   1   2   3   4
Kilometre 1 0 1 2 3 4 5 6
1 Centimetre

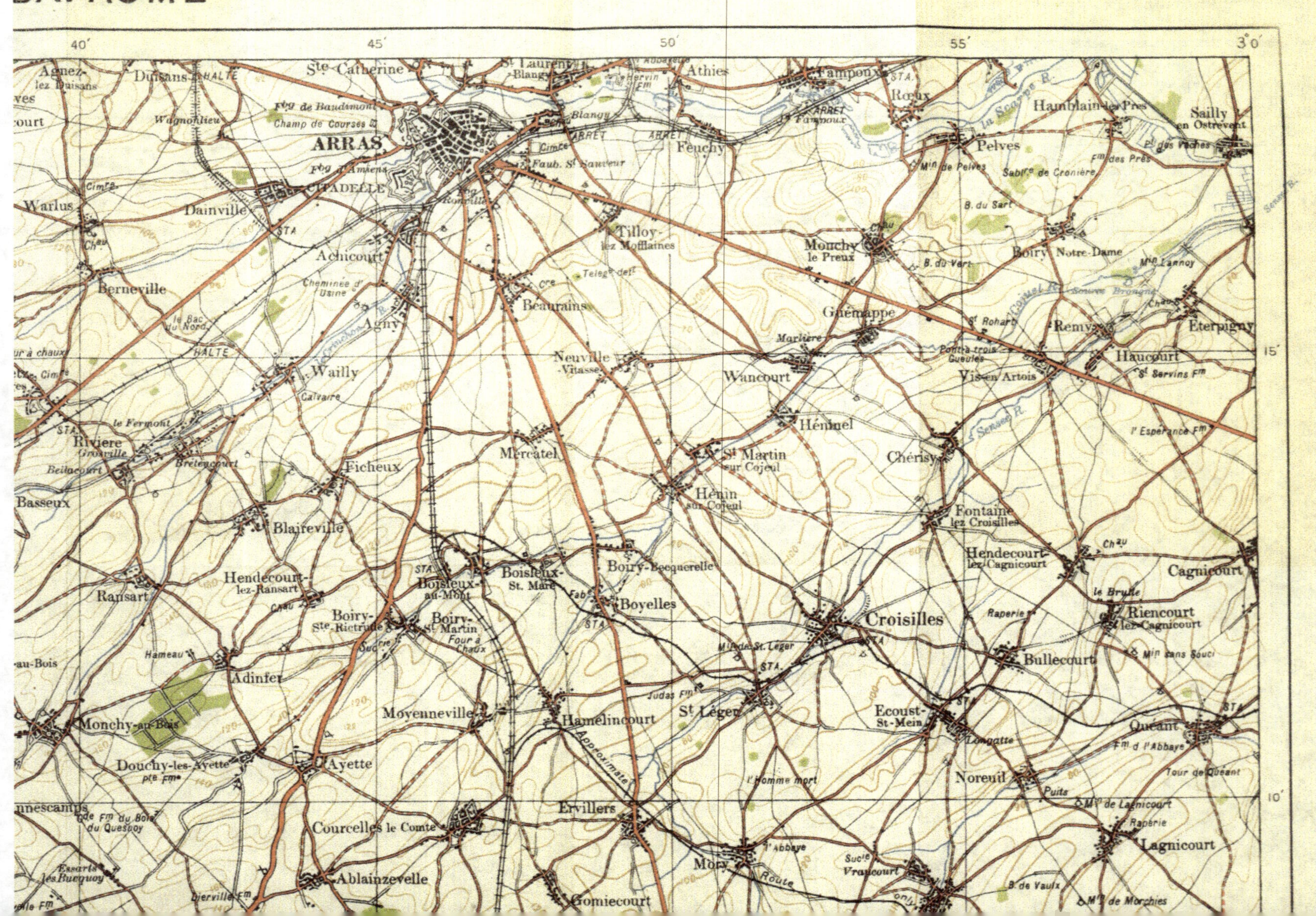

BAPAUME
40' 45' 50' 55' 3°0'
Agnez-lez-Duisans
Duisans
HALTE
Ste Catherine
St Laurent-Blangy
Athie
Fampoux
STA
Roux
Hamblain les Prés
Sailly en Ostrevent
Wagnonlieu
Fg de Baudimont
Champ de Courses
ARRAS
CITADELLE
Fg d'Amiens
Blangy
Cimée
ARRET
ARRET
Feuchy
Min de Pelves
Pelves
Sablé de Cronière
Fme des Prés
P des Vaches
B. du Sart
Warlus
Cimée
Dainville
STA
Achicourt
Tilloy-lez-Mofflaines
Telegre deff
Mouchy le Preux
B. du Vert
Boiry Notre-Dame
Mm Lannoy
Berneville
Cheminée d'Usine
Beaurains
Ghémappe
Marlière
St Rohart
Remy
Eterpigny
le Bac du Nord
Agny
Pont à trois Gueules
Haucourt
St Servins Fm
HALTE
Wailly
Calvaire
Neuville-Vitasse
Wancourt
Visen Artois
l'Espérance Fm
Rivière
Grouville
le Fermont
Mercatel
St Martin sur Cojeul
Héninel
Chérisy
Beltcourt
Brétencourt
Ficheux
Hénin sur Cojeul
Basseux
Blaireville
Boiry-Becquerelle
Fontaine lez Croisilles
Hendecourt lez Cagnicourt
Ch au
Hendecourt-lez-Ransart
STA
Boisleux-au-Mont
Boisleux-St. Marc
Cagnicourt
Ransart
Chau
Fab
le Brulle
Boiry-Ste.Rictrude
Boiry-St. Martin
Four à Chaux
Boyelles
Croisilles
STA
Raperie
Riencourt lez Cagnicourt
Hameau
Sucre
Mil de St Léger
STA
Min sans Souci
Adinfer
Judas Fm
St Léger
Bullecourt
Monchy-au-Bois
Moyenneville
Hamelincourt
Ecoust-St-Mein
STA
Quéant
STA
Douchy-les-Ayette
pte Fme
Ayette
Approximate
l'Homme mort
Longatte
Noreuil
Puits
Fm d l'Abbaye
Tour de Quéant
nnescamps
Fme du Bois du Quesnoy
Courcelles le Comte
Ervillers
Puits
Mil de Lagnicourt
Raperie
Essarts lès Bucquoy
Bierville Fm
Ablainzevelle
Mory
Route
Sucre Vraucourt
Gomiecourt
B. de Vaulx
Lagnicourt
Mil de Morchies
30'
15'
10'

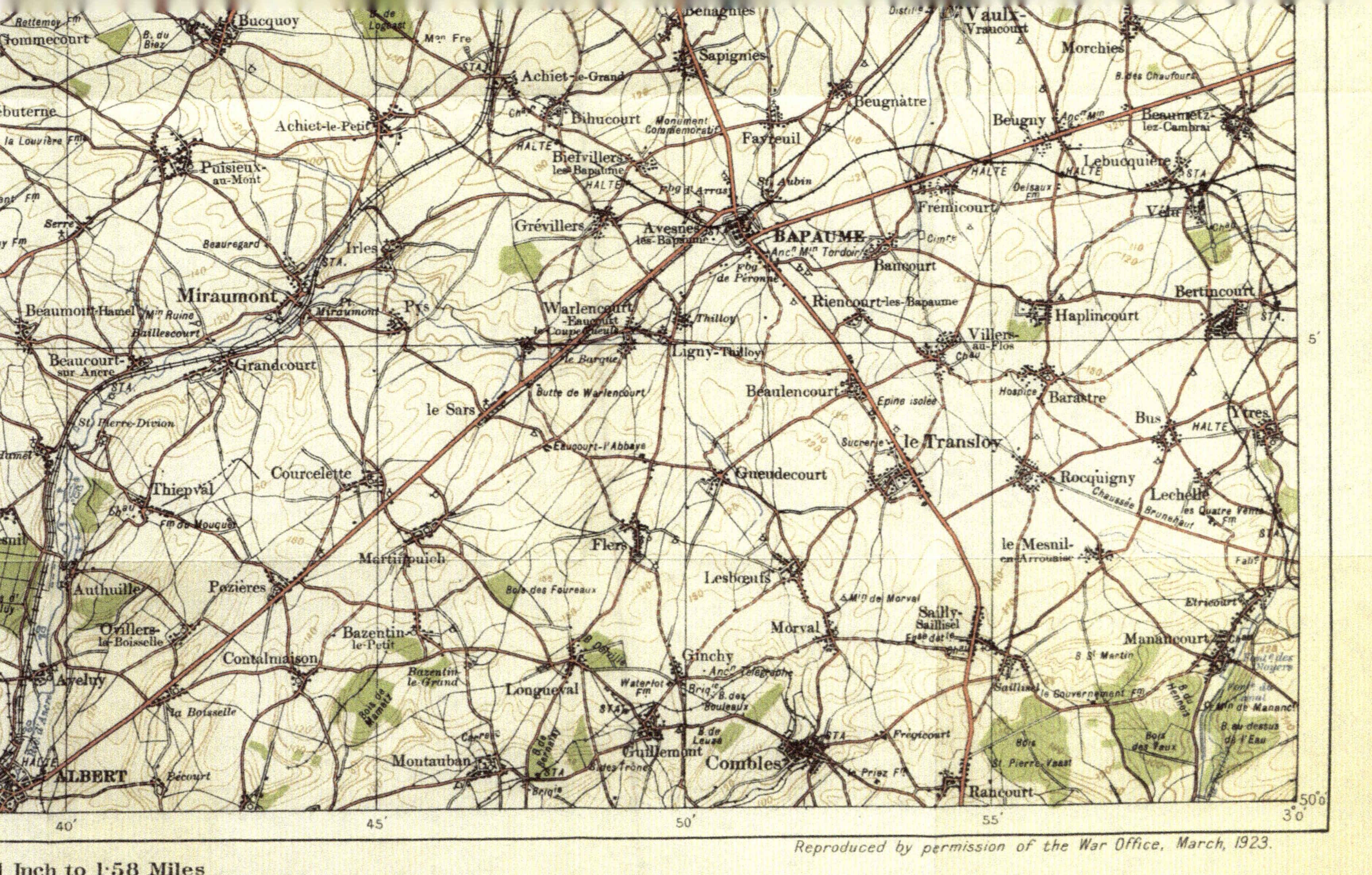
Gommecourt
Bucquoy
Behagnies
Vaulx-Vraucourt
Morchies
B. du Biez
Rettemoy Fm
B. de Logeast
Mon Fre
Sapignies
B. des Chaufours
thuterne
Achiet-le-Grand
Beugnâtre
Beugny
Beaumetz-lez-Cambrai
la Louvière Fm
Achiet-le-Petit
Bihucourt
Monument Commemoratif
Fayreuil
Ancn Mon
HALTE
Lebucquière
STA
ent Fm
Serre
Puisieux-au-Mont
Bietvillers-les-Bapaume
HALTE
Fbg d'Arras
St Aubin
HALTE
Frémicourt
Delsaux Fm
HALTE
Véle
Grévillers
Avesnes-les-Bapaume
BAPAUME
Ancn Mon Tordoir
O Cinte
y Fm
Beaumont-Hamel
Mon Ruine
Bazillescourt
Irles
STA
Miraumont
Miraumont
Pys
Warlencourt-Eaucourt
le Couperette
Thilloy
Fbg de Péronne
Bancourt
Riencourt-les-Bapaume
Villers-au-Flos
Chau
Haplincourt
STA.
Bertincourt
Beaucourt-sur-Ancre
STA
Grandcourt
le Barque
Ligny-Thilloy
Beaulencourt
Epine isolée
Hospice
Barastre
Bus
HALTE
Ytres
le Sars
Butte de Warlencourt
Sucrerie
le Transloy
St Pierre-Division
Eaucourt-l'Abbaye
Rocquigny
Chaussée
Lechelle
les Quatre Vents
Brunehaut
STA
nui
Courcelette
Thiepval
Chau
Fm du Mouquet
Gueudecourt
le Mesnil-en-Arrouaise
Fab
Flers
Authuille
Pozières
Martinpuich
Bois des Foureaux
Lesboeufs
Mon de Morval
Sailly-Saillisel
Etricourt
Manancourt
Chau
Ovillers-la-Boisselle
Bazentin-le-Petit
Morval
Esse de
B St Martin
Fond des Mogers
Contalmaison
Bazentin-le-Grand
Ginchy
Ancn Télégraphe
Saillisel le Gouvernement Fm
B. du
Aveluy
Longueval
Waterlot Fm
Brigie
B. des Bouleaux
Bois de Mametz
St Mon de Mananc
Bois des Vaux
la Boisselle
STA
B. de Leuze
B. de
STA
Frégicourt
Bois
St Pierre-Vaast
B. au-dessus de l'Eau
HALTE
ALBERT
Bécourt
Montauban
B. des Trônes
Guillemont
STA
Combles
B. Priez Fm
Rancourt
Reproduced by permission of the War Office, March, 1923.
1 Inch to 1·58 Miles
5   6   7   8   9   10 Miles
7   8   9   10   11   12   13   14   15 Kilometres
to 1 Kilometre

On October 13th General Sir C. Monro, commanding the Third Army, inspected the Battalion in the grounds of the chateau at Acheux, and in his address said : " it was always a pleasure to be able to say nice things to people, that he had very good accounts of the Battalion from everywhere, by reason of its bravery in action, its conduct in trenches and its smartness in billets ; he was sure that the Battalion would continue to live up to its traditions."

Some ten days later—on the 25th—several units of the VIIth Corps were inspected by His Majesty the King in Acheux, His Majesty being accompanied by M. Poincaré, the President of the French Republic ; the Battalion paraded at a strength of 15 officers and 624 other ranks on this occasion.

On October 29th the Battalion took over the sector on the immediate left which was just west of Beaumont-Hamel, and was the first unit to occupy an afterwards well-known place called " the White City."

Officers and men went in and out of this sector for the next two months and spent the greater part of their time in an uphill fight trying to keep the dug-outs and trenches habitable. The weather was very wet and the trenches in many cases became waist-deep in liquid mud ; the troops consequently lived in great discomfort in spite of every effort on the part of all ranks to ameliorate the conditions under which everybody existed. Only those who have had actual personal experience of the Somme mud can fully realize what the troops had to endure. But in spite of the terribly depressing circumstances the men were extraordinarily cheerful and the *morale* of the Battalion was high.

During the last three months of 1915 the casualties had fortunately been low, for the reinforcements had been singularly meagre ; in the months of October, November and December 1 officer—Captain R. W. Sutton—and 4 men had been killed, and 1 officer—Lieutenant L. G. Kettlewell—and 14 other ranks had been wounded, while the drafts sent out from home only totalled 6 officers and 63 men.

On December 26th, 1915, the effective strength of the Battalion was 26 officers and 813 non-commissioned officers and men.

At Christmas-time the following gracious message was received from His Majesty the King, ever mindful of the welfare of all ranks of his Army :—

"*Another Christmas finds all the resources of the Empire still engaged in war, and I desire to convey on my own behalf, and on behalf of the Queen, a heartfelt Christmas Greeting and our good wishes for the New Year, to all who, on Sea and Land, are upholding the honour of the British name.   In the officers and men of my Navy, on whom the security of the Empire depends, I repose, in common with all my subjects, a trust that is absolute.   On the officers and men of my Armies, whether now in France, in the East, or in other fields, I rely with an equal faith, confident that their devotion, their valour, and their self-sacrifice will, under God's guidance, lead to Victory and an honourable Peace.   There are many of their comrades, alas, in hospital, and to these brave fellows also, I desire, with the Queen, to express our deep gratitude and our earnest prayers for their recovery.*

"*Officers and men of the Navy and Army, another year is drawing to a close, as it began, in toil, bloodshed and suffering ; but I rejoice to know that the goal to which you are striving draws nearer into sight.*

"*May God bless you and all your undertakings.*

" GEORGE, R.I."

CHAPTER IV

1916

THE BATTLE OF THE SOMME.

DURING the concluding weeks of the year 1915 one event of great importance had taken place affecting the leadership of the British Forces in France, for in December Field-Marshal Sir John French relinquished the command which he had held since the beginning of hostilities, and was succeeded by General Sir Douglas Haig on the 19th of that month.

In his farewell address to the troops the Field-Marshal said :—

"*In relinquishing the command of the British Army in France, I wish to express to the officers, non-commissioned officers and men with whom I have been so closely associated during the last sixteen months, my heartfelt sorrow in parting with them before the campaign, in which we have been so long together, has been brought to a victorious conclusion. I have, however, the firmest conviction that such a glorious ending to their splendid and heroic efforts is not far distant, and I shall watch their progress towards this final goal with intense interest, but in the most confident hope.*

"*The success so far attained has been due to the indomitable spirit, dogged tenacity which knows no defeat, and the heroic courage so abundantly displayed by the rank and file of the splendid army which it will ever remain the pride and glory of my life to have commanded during over sixteen months of incessant fighting. Regulars and Territorials, Old Army and New Army, have ever shown these magnificent qualities in equal degree. From my heart I thank them all.*

"*At this sad moment of parting, my heart goes out to those who have received life-long injury from wounds, and I think with sorrow of that great and glorious host of my beloved comrades who have made the greatest sacrifice of all by laying down their lives for their country.*

" In saying goodbye to the British Army in France, I ask them once again to accept this expression of my deepest gratitude and heartfelt devotion towards them, and my earnest good wishes for the glorious future which I feel to be assured."

" In January, 1916, the Allies seemed in a favourable position for the campaign of the new year. They had considerably increased their strength in men and material. France had trained her 1916 and 1917 classes, but had not yet used them at the front ; Britain had by the Derby scheme and the Military Service Act provided, along with troops in home camps, a potential force nearly twice the size of that which she had in the field ; Italy had large numbers at her depots, and had not yet called up the greater part of her possible reserves ; Russia was only waiting for small arms to bring forward reinforcements as great as her field army, and her recruiting ground was still enormous. In Britain the manufacture of munitions was very different from the lean days a year before ; and in France, where trade union restrictions were wholly set aside, and the question of fatigue and the workers' capacity was really understood, the daily output exceeded even her past records. Germany and Austria were already pressed for men. In Germany the 1916 and part of the 1917 class were now included in the field forces ; Austria-Hungary had already used up most of the 1916 class, had called up the 1917 class as early as October, 1915, and had warned her 1918 class. Further, in both countries the incorporation of elderly men had been carried to lengths undreamed of among the Allies."*

On January 1st, 1916, the Battalion took over the sector on the immediate left of that which it had been holding in November and December of the previous year, and thus found itself back in the sector which it had taken over from the French in August, 1915—west of the village of Serre.

On the following day Captain R. M. Watson, D.S.O., vacated the adjutancy of the Battalion on appointment as Staff Captain 10th Infantry Brigade, and was succeeded as Adjutant by Lieutenant L. C. Byrne.

* Buchan, 1st edition, Vol. XIII, pp. 116, 117.

MINE CRATER BLOWN UP ON JULY 1st, 1916, BEAUMONT HAMEL.

GERMAN BARBED WIRE AND SHELL HOLES AFTER RAIN,
BEAUMONT HAMEL.

At the end of January the Brigade went back into rest billets, the Battalion being quartered in the village of Orville, where it spent a quiet month of training, this being practically the first real spell of rest it had had since arrival in France. Here an officers' battalion mess was started, thus enabling all the officers to get to know each other to an extent which was quite impossible with the company messes.

It was during this month that the Brigade lost its popular commander, Brigadier-General C. P. A. Hull, C.B., who was promoted to the command of the 56th (London) Division, and was succeeded by Brigadier-General C. A. Wilding, C.M.G.

At the beginning of March the 2nd Battalion Royal Dublin Fusiliers left Orville and moved to St. Amand, and then on the 17th it went into the line in front of Bienvillers and opposite Monchy-au-Bois, relieving a battalion of the 110th Infantry Brigade. In this line there was a particularly nasty salient on the Battalion front, while the reserve billets in Bienvillers were only about 2,000 yards behind the line. Considering its proximity to the enemy, the village had not been very badly damaged, and a considerable number of the civilian inhabitants were still living in the place.

On March 10th the Commanding Officer received the following letter :—" Mr. Redmond presents his compliments to the Commanding Officer of the 2nd Royal Dublin Fusiliers, and begs to say that he has forwarded to him a parcel of shamrocks for the men in his Battalion for St. Patrick's Day, with his very warmest good wishes for their success and good fortune."

By the end of this month the casualties sustained by the Battalion since it left Ypres amounted to 18 killed, 89 wounded, and 1 missing.

The following was the distribution of the officers of the Battalion on April 28th :—Lieutenant-Colonel H. W. Higginson, D.S.O., in command ; Major L. P. Walsh, 2nd-in-command ; Lieutenant H. G. Aylmer, adjutant ; Major J. Burke, quartermaster ; Captain C. St. L. Webb, transport officer ; Rev. Father F. S. Dowling, chaplain ; Captain C. H. Attenborough, R.A.M.C., in medical charge.

"*A*" *Company* : Captain T. Brady, Second-Lieutenants S. Craddock, C. B. Donovan, W. I. Black, R. H. Ingoldby, and G. S. T. Fenning.

"*B*" *Company* : Captain W. F. Jeffries, Second-Lieutenants S. J. Diamond, J. W. R. Morgan, A. W. Sainsbury, J. A. Tobin, and R. T. Dade.

"*C*" *Company* : Captain J. E. Long, Second-Lieutenants A. J. Franklin, F. T. A. Power, W. H. A. Damiano, and J. Gaffney.

"*D*" *Company* : Captain J. D. Glegg, Lieutenant R. M. Patterson, Second-Lieutenants R. McI. Stobart, W. T. Colyer, J. B. Moffat, and H. G. Killingley.

On April 30th Lieutenant-Colonel Higginson learnt that he had been appointed to the command of the 53rd Infantry Brigade, 18th Division, and left two days later, handing over command of the Battalion to Major Walsh. At the same time the 4th Division was taken out of the VIIth Corps (General Snow) and transferred to the VIIIth (General Hunter-Weston), and on May 1st the Battalion marched by way of Halloy, Mezerolles, Cramont, and Longvillers to Argenvillers, where and at Longvillers divisional training was carried on throughout the month.

On leaving the VIIth Corps, General Snow published the following farewell order to the 4th Division and its commander :—

"*On the occasion of your leaving the VIIth Corps, the Corps Commander wishes to express his regret at parting with such a good fighting body of men as those who compose the 4th Division. He wishes you and the division which he had the honour to bring out to this country the very best luck and further successes.*"

The VIIIth Corps, to which the Battalion now belonged, contained the 4th, 29th, 31st, and 48th Divisions, and, like the rest of the British Army, was preparing by a strong attack to take some of the pressure off the French at Verdun.

" The continued German pressure at Verdun, which had reached a high point in June, called insistently for an immediate allied attack at the western end of the line. With a fine spirit of comradeship, General Haig had placed himself and his armies at the absolute disposal of General Joffre, and was prepared to march them to Verdun, or anywhere else where he could best render assistance. . . . The southern end of the whole British line was held by the Fourth Army, commanded by General Rawlinson. . . . His army consisted of five corps, each of which included from three to four divisions, so that his infantry numbered about 200,000 men, many of whom were veterans, so far as a man may live to be a veteran amid the slaughter of such a campaign. The Corps, counting from the junction with the French, were the XIIIth (Congreve), XVth (Horne), IIIrd (Pulteney), Xth (Morland), and VIIIth (Hunter-Weston). North of Rawlinson's Fourth Army, and touching it at the village of Hébuterne, was Allenby's Third Army, of which one single corps, the VIIth (Snow) was engaged in the battle. This added three divisions, or about 30,000 infantry, to the numbers quoted above.

" It had taken months to get the troops into position, to accumulate the guns, and to make the enormous preparations which such a battle must entail. . . . Every operation of the attack was practised on similar ground behind the lines. New railheads were made, huge sidings constructed, and great dumps accumulated. . . . Assembly trenches had to be dug, double communication trenches had to be placed in parallel lines, one taking the up-traffic, the other the down, water supplies, bomb shelters, staff dug-outs, poison gas arrangements, tunnels and mines—there was no end to the work of the sappers. . . . The question of ammunition supply had assumed incredible proportions. For the needs of one single corps, 46 miles of motor lorries were engaged in bringing up the shells. However, by the end of June all was in place and ready."*

The Battalion remained during the whole of May in the neighbourhood of Argenvillers practising the attack and preparing for the battle which was imminent, and then, leaving this place on June 4th, it marched in heavy rain by way of Bernaville, Authie,

* Conan Doyle, Vol. III, p. 33 *et seq*.

and Bertrancourt to trenches in the neighbourhood of Mailly-Maillet and Auchonvillers. The bombardment of the enemy trenches began on June 24th, and the gun fire was very heavy on both sides, especially at night ; on the 25th the Battalion, which was that day out of the line in Mailly-Maillet, had to move out into the fields west of that village, as so many German shells were then falling in reply to the intense bombardment by the British artillery.

On the evening of the 26th the 2nd Royal Dublin Fusiliers moved by a circuitous route to Beaussart, the move being completed by 11.20 p.m., and here bivouacked in some orchards, and the following day was passed in completing arrangements for the coming battle.

The attack had been planned to open on June 28th, but the weather had been so very wet and the ground had become so exceptionally muddy, that the attack was postponed until July 1st, which broke calm and warm with a gentle south-west breeze, and at 10.15 the previous night the Battalion had moved up to the trenches and occupied the assembly position which had there been prepared.

At the time that the Battle of the Somme of this year opened the VIIIth Corps was occupying a front which extended from Hébuterne in the north, where it joined on to the 56th Division, down to a point just north of the Ancre, and it faced the very strong German positions at Serre in the north and at Beaumont-Hamel in the centre. Here the position was particularly difficult, for it contained huge quarries and excavations in which masses of German troops could and did remain concealed, and in almost complete safety, ready to sally forth when their services were required. Here, as may easily be realized, the heavy British bombardment had done comparatively little damage, and the enemy had not been greatly injured either materially or in *morale*.

The order of battle of the VIIIth Corps was as follows : the 48th Division was in reserve except that two battalions of one of its brigades held a defensive line about a mile south of Hébuterne, while two others were attached during the battle to the 4th Division, commanded since the previous September by Major-General the Hon. W. Lambton, C.B., C.M.G., C.V.O.

Immediately south of the defensive line above mentioned was the 31st Division, which had Serre for its objective. South of this, and opposite to Beaumont-Hamel, was the 4th, and south of this again was the 29th Division, lately back from the Dardanelles, and containing, in its 86th Brigade, the 1st Battalion Royal Dublin Fusiliers.

Of the 10th and 12th Brigades two battalions of each advanced at 9 a.m. ; on the right the 2nd Royal Dublin Fusiliers and 2nd Seaforth Highlanders of the 10th, and on the left the 2nd Battalion Essex and 1st Battalion King's Royal Lancaster Regiment of the 12th Brigade.

At 9 a.m. then the Battalion left its assembly trenches and advanced in the following formation :—" A," " B " and " C " Companies in four lines of platoons, each platoon in sections in artillery formation, the lines a hundred yards apart, and with " D " Company in reserve 200 yards in rear of the fourth line ; this company was in diamond formation, each platoon being in column of fours.

Immediately after leaving their assembly trenches each company came under a heavy enfilade fire from machine guns in Beaumont-Hamel, even the reserve company sustaining casualties as it left Young Street. At 9.5 a.m. the order was received to " stand fast," followed immediately by the message :—" Your battalion not to go beyond English front line trenches till further orders." Two runners were at once sent after each of the advancing companies, and Major Walsh personally went to " A " Company and was just in time to catch a sergeant who was slightly in rear of the 4th line and who at once doubled forward with the message. But it was only delivered just in time to stop the advance at the first line trenches.

In the meantime the Battalion had been losing heavily, not only from the machine-gun fire from Beaumont-Hamel, but from rifles and machine guns from the German first line trenches in the

immediate front, and also from shell-fire. It had been impossible
to halt all the platoons and some had got into the ground between
the enemy and the British trenches ; all these without exception
became casualties, including five officers.

At 12 noon the following order was received :—" You will
attack the German trenches and consolidate line from Point 86 to 88
inclusive. The Seaforth Highlanders are attacking north of the
Redan.   Point 59 is held by our own troops.   The 29th Division is
attacking Beaumont-Hamel at 12.30.   Take care of your left flank
as there are still some Germans in position opposite the Redan."
It was, however, found impossible by this time to collect more than
sixty men, for during the advance, out of the 23 officers and 480
men who left the assembly trenches that morning, 14 officers and
311 non-commissioned officers and men had already become
casualties, while since the advance had been made in lines of
sections on a front of 400 yards, the survivors had become much
scattered and mixed up with those troops remaining out of the
11th Brigade,

The Commanding Officer of the Battalion was now therefore
directed to collect all the men he could and withdraw them into their
original assembly trenches, and this was accordingly done.

" Late in the afternoon the assault had definitely failed, and
the remainder were back in their own front trenches, which had now
to be organised against the very possible counter-attack. Only
two battalions of the division remained intact and the losses included
one brigadier and seven battalion commanders killed and wounded.
For a long time a portion of the enemy's trench was held by mixed
units, but it was of no value when detached from the rest and was
abandoned in the evening.   From the afternoon onwards no possible
course save defence was open to General Lambton. . . .

" It must have been with a heavy heart that General Hunter-
Weston realised, with the approach of night, that each of his divisions
had met with such losses that the renewal of the attack was
impossible. He, his divisional commanders, his officers and his
men had done both in their dispositions and in their subsequent

IN BEAUMONT HAMEL, NOVEMBER, 1916.

AT A POST IN A FRONT LINE TRENCH.

actions everything which wise leaders and brave soldiers could possibly accomplish. If a criticism could be advanced it would be that the attack was urged with such determined valour that it would not take ' No ' until long after ' No ' was the inevitable answer."*

During the morning of July 2nd the Battalion was reorganized and at 2 p.m. was distributed—1 company in a trench immediately in front of Tenderloin Street, 1 company in Young Street and the remaining 2 in the Sunken Road, and then at 5 p.m. on the 3rd was withdrawn to billets in Mailly-Maillet.

The following were the casualties sustained by the Battalion on July 1st :—Killed—Second-Lieutenants J. W. R. Morgan and R. H. Ingoldby ; wounded—Major L. P. Walsh, Captains J. O. W. Shine, W. F. Jeffries and T. Brady, Lieutenants W. T. Colyer and H. E. Andrews, Second-Lieutenants W. H. A. Damiano, T. W. H. Mason, A. W. Sainsbury, A. E. Mulholland, A. J. Franklin and J. C. J. Chadwick, while, as already stated, 311 non-commissioned officers and men were killed, wounded and missing.

On July 2nd Second-Lieutenant Damiano died of his wounds, as did two days later Major Walsh ; this was a very gallant and capable officer, who had joined the 3rd Battalion of the Regiment at the beginning of the war, travelling from America at his own expense to do so, and being given a commission as captain in the 3rd Battalion, obtaining in 1916 a regular commission as captain in the 2nd Battalion.

Lieutenant-Colonel H. M. Cliff assumed command of the Battalion for a few days, being on July 13th relieved by Major R. G. B. Jeffreys.

At 6 a.m. on the 20th the Battalion received orders to leave the Somme area, and marched early the same afternoon by Beauval to Candas where it entrained and proceeded to Proven, being here accommodated for some three or four days in a camp about three miles east of Poperinghe, where H.R.H. the Prince of Wales was camp commandant. Proven was left again by train on the 27th, and the Battalion was sent to the neighbourhood of the Yser canal where it became the right reserve battalion.

* Conan Doyle, Vol. III, p. 52 *et seq.*   F

Previous to leaving the Somme area the following gracious message from the King had been promulgated in Army Orders by General Sir Douglas Haig :—

" Please convey to the Army under your command my sincere congratulations on the results achieved in the recent fighting. I am proud of my troops. None could have fought more bravely."

The situation in the new area to which the Battalion had now been transferred is described in the war diary as " quiet," but casualties appear none the less to have at once occurred ; thus on the early morning of August 3rd Second-Lieutenant Helby was on patrol and he and one of his party were wounded by bomb splinters, Helby dying during the day of his wounds. During this month, when the Battalion was not in the trenches, the headquarters were first at Elverdinghe Chateau and then at Trois Tours Chateau, from where training was carried out ; but in the last week of the month the Battalion made another move, this time by train to Winnipeg Camp and while here occupied trenches opposite Hill 60.

During August His Majesty the King paid a visit to his Army in France and Flanders and expressed himself as pleased with what he was able to see of the troops in the VIIIth Corps area, and on the 15th he caused the following expression of his satisfaction to be issued :—

" Officers, non-commissioned officers and men.

" It has been a great pleasure and satisfaction to me to be with my armies during the past week. I have been able to judge for myself of their splendid condition for war and of the spirit of cheerful confidence which animates all ranks united in loyal co-operation to their Chiefs and to one another.

" Since my last visit to the front there has been almost uninterrupted fighting on parts of our line. The offensive recently begun has since been resolutely maintained by day and by night. I have had opportunities of visiting some of the scenes of the later desperate struggles, and of appreciating to a small extent the demands made upon your courage and physical endurance in order to assail and capture positions prepared during the past two years and stoutly defended to the last.

" I have realised, not only the splendid work which has been done in immediate touch with the enemy—in the air, underground, as well as on the ground—but also the vast organisations behind the fighting line, honourable alike to the genius of the initiators and to the heart and hand of the workers. Everywhere there is proof that all, men and women, are playing their part, and I rejoice to think their noble efforts are being heartily seconded by all classes at home.

" The happy relations maintained by my Armies and those of our French Allies are equally noticeable between my troops and the inhabitants of the districts in which they are quartered, and from whom they have received a cordial welcome ever since their first arrival in France.

" Do not think that I and your fellow-countrymen forget the heavy sacrifices which the Armies have made and the bravery and endurance they have displayed during the past two years of bitter conflict. These sacrifices have not been in vain ; the arms of the Allies will never be laid down until our cause has triumphed.

" I return home more than ever proud of you.

" May God guide you to victory.

" GEORGE R.I."

The following were the officers present with the 2nd Battalion Royal Dublin Fusiliers at the end of August :—Lieutenant-Colonel R. G. B. Jeffreys ; Captains J. G. F. Napier-Martin, C. St. L. Webb, W. A. Redmond, J. D. Glegg, adjutant, R. M. Patterson, H. G. Aylmer and R. MacI. Stobart ; Lieutenants L. A. King, L. G. Kettlewell, J. A. Noblett, J. B. Moffat, J. Gaffney and S. Craddock ; Second-Lieutenants T. F. Handyside, J. B. Sheehan, E. C. B. Dillon, W. G. Scott, W. N. Henchy, G. S. T. Fenning, W. I. Black, H. G. Killingley, F. T. A. Power, W. Pedlow, L. G. Doran, C. B. Donovan, B. P. Glancy, H. J. Lemass, W. V. Beaumont, A. R. Henry, P. N. Gordon, W. H. Hynes, R. G. Bourke, R. F. Nalder and H. D. Gibson ; Hon. Major and Quartermaster J. Burke ; Lieutenant H. I. G. Rutherford, R.A.M.C. and the Rev. E. Dowling. At the same date the effective strength of " other ranks " was 921.

During September and the first half of the month following the Battalion was constantly moved about, the Brigade seeking suitable centres for brigade and battalion training. On September 1st the Dublin Fusiliers were in trenches opposite the " Dump," but on relief that night by the 3rd Battalion Australian Infantry, they marched to Ypres, there entrained for Brandhoek and thence marched to Erie Camp, half-a-mile south of the Ypres—Poperinghe road ; this camp was left on the 4th the destination being Poperinghe, where the stay was only a very short one, the Battalion moving again on the 9th and marching by St. Jan-Ter-Biezen — Watou—Winnezeele—Haegdeone—Wemaers—Cappel and Noordpeene to Tatinghem, where the regiment arrived on the afternoon of the 10th, having covered a good 20 miles, the men swinging along splendidly and marching at the rate of 7 miles in $2\frac{1}{4}$ hours ; the weather though fine was rather warm for marching.

Here it was hoped that the Battalion might have remained quietly for at least a month, and all ranks prepared to settle down to the curriculum of training laid down by the Second Army Central School of Instruction ; but on September 16th orders were received that all parades, etc., were to be cancelled and that the Battalion was to stand by ready to march off at the shortest possible notice, and later in the day it was directed that entraining would take place on the morrow.

On the 17th then Tatinghem was left and the Battalion marched to and boarded the train at St. Omer ; that night was passed in very crowded billets at Rainneville and company training was resumed ; but on the 24th the 2nd Royal Dublin Fusiliers marched to Cordie, on the 25th to Sailly-le-Sec, several hostile aeroplanes passing over the column *en route ;* and by the end of the month they were in Daours, where battalion training was carried out, and brigade and divisional programmes were arranged, but frequently postponed or wholly abandoned, and training was conducted under unusual difficulties, while the shadow of a fresh move was upon all ranks.

On October 7th the Brigade left Daours marching by Corbie,

THE CATHEDRAL AT YPRES, NOVEMBER, 1916.

VIEW OF YPRES, TAKEN AT THE END OF 1916.

Mericourt, and Ville-sur-Ancre to Meaulte, halting for dinner at Mericourt ; the night was passed in billets in Meaulte and next day the Battalion moved to Mansel Camp near Carnoy, passing over a very desolate country, harassed by many recent bombardments. The men were accommodated in tents and, owing to the late heavy rains, the camping ground was very muddy—as was only to be expected in the Somme area, to which the 2nd Battalion Royal Dublin Fusiliers had now returned.

At 7 a.m. on the 9th the Battalion moved off at intervals of 100 yards between companies to a place just east of Trones Wood and remained here during the day, the first line transport staying near Carnoy ; here the 10th Brigade was thus distributed : the 1st Royal Warwickshire Regiment and 1st Royal Irish Fusiliers were in the front line on the right and left respectively, the Royal Dublin Fusiliers were in support, and the Seaforth Highlanders in reserve. The night passed quietly enough, but on the next day, and during the whole of the night of October 10th–11th, the enemy shelled the front and support lines heavily with 5.9 and 8-in. guns, our own artillery replying with great vigour ; the Battalion war diary now describes the events of the 12th and this is supplemented by a report on the operations drawn up by Lieutenant-Colonel Jeffreys.

" The 1st Royal Irish Fusiliers and 1st Royal Warwickshire Regiment are going to carry out an attack on the enemy front line on the morrow.   We are sending up two companies, ' A ' and ' B,' to occupy our front line when the two regiments have moved off. The French are attacking on our right.   The 56th Division have made two attacks on this line, but have been driven back owing to the fire from the Gun Pits and a Strong Point in the centre of our line."   So far the Battalion diary and we now take up Colonel Jeffreys' report.

" On October 12th the 10th Brigade was ordered to attack the line Dewdrop—Gun Pits—Hazy Trench.   The Battalion was distributed as follows :—1 company, Fluff Trench, 3 companies T.9.a.   The battalions attacking were 1st Warwicks on right, 1st Royal Irish Rifles on left, 2nd Royal Dublin Fusiliers supported

Warwicks, 2nd Seaforths, reserve. At 2.5. p.m. ' A ' Company, Lieutenant Gaffney, moved to German Trench vacated by 1st Royal Warwickshire Regiment; ' B ' Company, Captain Byrne, moved up to Fluff Trench. From information received that Warwicks were in need of support, I ordered ' A ' Company up; they got up without many casualties. They eventually pushed forwarded and assisted to dig a new trench, now called Antelope Trench. They met with a certain amount of resistance. ' B ' Company reinforced the Royal Irish Fusiliers, but were not heavily engaged with the enemy.

" Information was received from O.C. Warwicks that Strong Point was still in possession of the enemy and I ordered ' A ' Company to attack the position from the south-west; my message, however, never reached its destination. I later got the order from the Brigade to find out if Dewdrop and Hazy Trenches were occupied by the enemy and endeavoured to do so, but communication was impossible. My Battalion was eventually withdrawn from the front line that night."

In the course of this day's operations 8 men of the Battalion were killed, Second-Lieutenants P. N. Gordon and R. G. Bourke and 54 non-commissioned officers and men were wounded, and 4 men were missing.

This month, unfortunately, was to be one of recurring casualties for the 2nd Battalion Royal Dublin Fusiliers. There was heavy enemy shelling all day on the 13th; that evening the Battalion went up to the front line again, relieving the Warwicks who went into reserve, and having the Seaforths on the left and the Royal Irish Fusiliers as the supporting battalion, and during the course of the relief Second-Lieutenant E. C. B. Dillon was killed by a 77 mm. shell on the Les Bœufs—Morval Ridge and there were four other casualties. During this night the Seaforth Highlanders, with 50 men of the Dublin Fusiliers, were ordered to take the Gun Pits and Strong Point by a bayonet charge; the enemy did not immediately discover what was taking place, but then opened a heavy machine-gun and rifle fire, putting down a barrage within half a minute and it was practically impossible to get through it. Some of the Highlanders did get into the Gun Pits, but they were so few that they were almost at once driven out by a strong bombing attack, and of the 50 men of the Battalion who were engaged 16 were wounded.

The 14th and 15th and the night that followed were tolerably quiet except for the continuous bombardment by the English heavy artillery, but between October 14th and 22nd 11 men were killed and 19 wounded.

On this last-named date Colonel Jeffreys received orders that the Battalion was to go up next day to the trenches, being attached for the operations arranged for that day to the 11th Brigade, and at 4.10 p.m. the Battalion marched to the assembly positions, which were in those known as Burnaby, Foggy, Muggy Trenches, and in a fourth joining up Foggy Trench and Andrews Post and called later New Trench.

Everything was ready for the advance soon after dawn on the 23rd, but no movement took place until the afternoon, when the Battalion was disposed as follows:—Leading companies, " D," Lieutenant Moffat, on the right in New Trench; " C " Company, Captain Patterson, on the left in the south end of Burnaby Trench and north end of Foggy Trench; supporting companies—" A," Lieutenant Gaffney, in left of German Trench; " B " Company, Lieutenant Byrne, in south end of Foggy Trench, supporting " D " and " C " Companies respectively.

The objective having been carefully pointed out, at 2.30 p.m. the Battalion " went over the top " in four waves and doubled up to the English artillery barrage; the companies on the right, being some distance in rear of those on the left, had some way to double before they were able to get up in line, but the whole got forward in sufficient time to escape the enemy barrage. No opposition was encountered until within 10 yards of the Gun Pits, when a heavy machine-gun and rifle-fire was met, compelling the leading lines to lie down. They, however, managed to crawl forward and bomb the Gun Pits, and eventually got into them, where very desperate hand-to-hand fighting ensued. The Gun Pits were strongly built and were armed with four machine guns ; three of these were here destroyed, and one is now in possession of the Battalion.

" C " Company, led by Captain Patterson, went right through and over the Gun Pits and on to Strong Point, which was greatly damaged by shell-fire ; but trifling opposition was here met with, and " C " Company, pushing on, established a line some little distance beyond the Gun Pits, while " A " and " B " Companies, remaining in the Gun Pits, thoroughly cleared them and all surrounding shell-holes, when " A " went up to the front line, followed almost at once by " D," leaving 20 men to hold the Gun Pits.

" D " Company on the right had somewhat lost connection, but a platoon on the right regained touch with the Rifle Brigade, while the remaining three platoons, changing their direction quarter left, made for the Gun Pits also, these being the only conspicuous objective here discernible. This movement caused a gap in the line of about a hundred yards, led to considerable congestion at this point, and engendered many casualties ; but probably the majority of these were sustained in the hand-to-hand fighting above mentioned or were the result of the heavy machine-gun fire.

The whole of the Gun Pits, Strong Point, and the line of shell-holes were now held and consolidated, and third lines dug and joined up in front of the Strong Point. The line joining these was dug as a defensive flank trench, and for protection from an enemy machine gun which was causing much trouble from south of Dewdrop Trench. These lines were also joined up into the Gun Pits, and the whole system of lines was already some three feet deep when eventually handed over by the Battalion on relief.

The Dublin Fusiliers held on here until very early on the morning of the 25th, when a Battalion of the 55th Division took their place, and by the 27th they were back again in billets at Carnoy with their own brigade.

The losses sustained by the Battalion on the 23rd were as follows :—

Killed : Second-Lieutenants L. G. Doran, H. G. Killingley, and H. J. Lemass, and 14 non-commissioned officers and men.

Wounded : Lieutenant J. Gaffney, Second-Lieutenants B. P. Glancy, R. F. Nalder, W. H. Hynes, H. D. Gibson, and 124 other ranks.

Missing : 36 non-commissioned officers and men.

It will be seen that the Battalion had had considerable losses during this month in both officers and men, and against these only 9 officers and 26 other ranks had joined in the same period.

The morning of the 28th was spent in reorganizing the Battalion, and in the afternoon there was a parade, when both the brigade and divisional commanders thanked officers and men for the good work they had done in the attack on the Gun Pits, while General Lambton further conveyed to all ranks the thanks of General the Earl of Cavan, commanding XIVth Corps, for the assistance they had given to a brigade of one of his divisions. On this day orders were received for a further move, and, starting about mid-day on the 30th, the Battalion proceeded by rail and road, via St. Sauvent and Arraines, to Huppy, which was reached the same evening ; and then, on November 2nd, after a couple of hours' march, very cramped billets were reached, Headquarters and " C " Company being quartered in Rogeant and the remaining three companies in Tœufels ; here training was commenced and carried out in very damp, cold weather.

On the 8th there was a presentation of Military Medals to twelve men of the 4th Division by Lieutenant-General Du Cane, commanding the XVth Corps, when Sergeants Gaffney, Wallace, and Carroll of the Battalion were handed the medals they had won ; and Major-General Lambton, who was going home on leave, took the opportunity of bidding the Battalion good-bye, for, as he stated, it would have left the 4th Division prior to his return.

The 2nd Battalion Royal Dublin Fusiliers was now to leave the Brigade in which it had served from the opening of the war, and was transferred to the 48th Brigade of the 16th (Irish) Division, the other battalions composing the 48th Brigade being the 7th Battalion Royal Irish Rifles, and the 8th and 9th Battalions Royal Dublin Fusiliers.

Marching on the morning of November 15th to Gamaches, the

Battalion there entrained to Bailleul and marched thence to
Meteren, which was reached on the night of the 16th, and here the
companies were comfortably billeted in farms. On the 22nd the
Dublin Fusiliers marched to Locre and became divisional reserve,
" A " and " B " Companies being in Birr Barracks, " C " and " D "
in Doncaster Huts.   Here two days later the Battalion was inspected
by Brigadier-General Ramsay, D.S.O., commanding 48th Infantry
Brigade, and three days later, again, all ranks were proud to learn
that No. 11213 Sergeant Robert Downie, of " B " Company 2nd
Battalion Royal Dublin Fusiliers, had been awarded the Victoria
Cross for gallantry in action on October 23rd. The following is
from the *London Gazette* of November 25th, 1916 :—

" For most conspicuous bravery and devotion to duty in attack.

*" When most of the officers had become casualties, this non-
commissioned officer, utterly regardless of personal danger, moved about
under heavy fire and reorganised the attack, which had been temporarily
checked. At the critical moment he rushed forward alone, shouting,
' Come on, the Dubs !'*

*" This stirring appeal met with immediate response, and the line
rushed forward at his call.*

*" Sergeant Downie accounted for several of the enemy, and, in
addition, captured a machine gun, killing the team. Though wounded
early in the fight, he remained with his company, and gave valuable
assistance whilst the position was being consolidated.*

*" It was owing to Sergeant Downie's courage and initiative that
this important position, which had resisted four or five previous attacks,
was won."*

On the 28th the Battalion relieved the 8th Royal Dublin
Fusiliers in the front line, taking over the left sub-sector east of
Kemmel. These trenches were in a bad state, much damaged by
age and weather ; the enemy was fairly quiet here, but the days and
nights were very cold and the men suffered a good deal, especially
the young soldiers who, to the number of over 300, had in this
month joined the service companies in drafts from home.

The effective strength of the Battalion at the beginning of

December was 32 officers and 957 non-commissioned officers and men.

Here, then, about Kemmel, the Battalion spent the remaining days of the year 1916, cheered during the month by a visit from the Commander-in-Chief, who inspected the Regiment and expressed his entire satisfaction at the general " turn-out."

On December 18th Major-General W. B. Hickie issued the following inspiring message to the 16th (Irish) Division :—

" *To-day is the anniversary of the landing of the 16th Division in France.   The Divisional Commander wishes to express his appreciation of the spirit which has been shown by all ranks during the past year.*

" *He feels that the Division has earned the right to adopt the motto which was granted by the King of France to the Irish Brigade which served in this country for a hundred years—' Everywhere and always faithful.'*

" *With the record of the past, with the memory of our gallant dead, with this motto to live up to, and with our trust in God, we can face the future with confidence.*

" *God save the King.*"

His Majesty King George again showed at this season, as on so very many occasions during the course of the war, that his soldiers were ever in his thoughts, as witness the gracious messages he sent them through their commander :—

" *No. 1.   I send you, my sailors and soldiers, hearty good wishes for Christmas and the New Year.   My grateful thoughts are ever with you for victories gained, for hardships endured, and for your unfailing cheerfulness.   Another Christmas has come round and we are still at war, but the Empire, confident in you, remains determined to win.*

" *May God bless and protect you !*"   " GEORGE, R.I."

" *No. 2.   At this Christmastide the Queen and I are thinking more than ever of the sick and wounded among my sailors and soldiers. From our hearts we wish them strength to bear their sufferings, speedy restoration to health, a peaceful Christmas, and many happier years to come.*   " GEORGE R.I."

CHAPTER V

1917

BATTLES OF MESSINES AND THIRD YPRES

" In the latter days of 1916 and the beginning of 1917 the British Army, which had in little more than two years expanded from seven divisions to over fifty, took over an increased line.   The movement began about Christmas time, and early in the New Year Rawlinson's Fourth Army, side-stepping always to the south, had covered the whole of the French position occupied during the Somme fighting, had crossed the Somme, and had established its right flank at a place near Roye. The total front was increased to 120 miles."*
" This alteration," wrote Field-Marshal Sir Douglas Haig in his despatch of May 31st of this year, " entailed the maintenance by the British forces of an exceptionally active front, and, combined with the continued activity maintained throughout the winter, interfered to no small extent with my arrangements for reliefs. The training of the troops had consequently to be restricted to such limited opportunities as circumstances from time to time permitted."

December had been wet, cold, and misty, and with the opening of the New Year came a period of bitter frost, varied by snowstorms, which tried sorely the endurance of the men in the trenches.

A general plan of campaign to be pursued by the Allied Armies during 1917 had been agreed on by a conference of military representatives in November, 1916, the plan comprising a series of offensives on all fronts, so timed as to assist each other by depriving the enemy of the power of weakening any one of his fronts in order to reinforce another. So far as the British Army was concerned, Sir Douglas Haig proposed to direct his initial efforts against the enemy troops occupying the salient between the Rivers Scarpe and

* Conan Doyle, Vol. IV, p. 1.

Ancre, into which they had been pressed as the result of the battles of the Somme ; it was intended to attack both shoulders of this salient simultaneously, the Fifth Army operating on the Ancre front, while the Third attacked from the north-west about Arras. So soon as this salient was " pinched off " as a result of the success of these operations, it was then proposed to transfer the offensive to another part of the front—to Flanders, where, and especially about the Ypres salient, the position was not altogether satisfactory.

The British front here was wholly overlooked by the enemy ; its defence involved much strain on the troops occupying the trenches, while their maintenance in case of serious attack was bound to be costly. It was evident that the British positions here would be greatly improved by the capture of the Messines— Wytschaete Ridge and of the high ground extending thence north-east for some 7 miles, and which thence trends north through Brood-seinde and Passchendaele. Certain preparations had already been put in hand, reinforcements had reached the British Army, while the new levies had abundantly proved their worth. " I therefore hoped," wrote the Field-Marshal in his despatch of December 25th, " after completing my spring offensive further south, to be able to develop this Flanders attack without great delay, and to strike hard in the north before the enemy realised that the attack in the south would not be pressed further."

For the first five months of the year, then, the 16th Division, and more particularly the 48th Brigade, held the line opposite Wytschaete, and while many raids were during that time here carried out upon the enemy's trenches, there were no major operations of any kind. An account may, however, usefully be given of one of these raids in which the 2nd Battalion Royal Dublin Fusiliers took part during that period.

Thus, about the beginning of the second week in January, a raid was planned upon a certain section of the German front line, to be carried out by 3 officers and 55 non-commissioned officers and men of the Battalion ; Captain G. A. Stoney was in command, and had with him Second-Lieutenants Beaumont and Alexander and an officer of the 48th Trench Mortar Battery. The men were divided into four officer-parties, and in order to ensure the element of surprise there was no artillery preparation, but the Vierstraat Group R.A. was directed to put down a barrage, if necessary, to cover the withdrawal of the infantry. The weather was wet and cold, and the condition of the trenches was deplorable.

Zero hour was fixed at 5.30 p.m. on the 11th, and the raiding party, which had been accommodated at Siege Farm, moved up to the trenches, and at 5.15 everything was ready, when two minutes later an intense rifle and machine-gun fire broke out from the enemy trenches, being concentrated exactly on those places whence the different parties were to have emerged. Five minutes later shells from trench mortars and rifle grenades began to fall on the trenches, and heavy bursts of machine-gun fire continued for some time. It was now evident that the enemy had by some means obtained very accurate information of what was in view, and the order to " stand fast " was first given, but later the enterprise was wholly abandoned.

None of the raiding party—the men were all volunteers—knew the place of attack, nor did they know the day or hour until the actual morning of the raid, and it was thought at the time that the inhabitants of Siege Farm had had something to do with " giving away " the scheme. There were numerous wires running all round the farm, and many of these had been tested, and no reply could be obtained through some of them.

The party did not sustain any casualties.

The enemy also occasionally showed an offensive spirit : thus on the evening of March 8th, when the Battalion was in reserve training, a message was received from the headquarters of the Brigade that the enemy had broken into the trenches held by the right battalion—the 7th Battalion Royal Irish Rifles. " B " and " D " Companies were sent off at once to reinforce, under command of Major Smithwick. It had been feared that the enemy would destroy a mine shaft which at this time was being sunk, but this was saved by the fine resistance offered by the miners, mostly men belonging to the 2nd Battalion Royal Dublin Fusiliers.   On arrival at the point threatened, " B " Company took over part of the front line, and held it until relieved by a company of the 8th Dublins, while " D " furnished working parties to repair the damage done.

On March 31st the Battalion left the neighbourhood of Kemmel for the training area at Zouafques, moving via Locre, Bailleul, Strazeele, Borre, and Hazebrouck, where the night was spent, and arriving at its destination on the afternoon of April 2nd.

While quartered in the Kemmel area during the months of January, February, and March, 25 officers and 65 other ranks had joined the Battalion ; while 8—including Second-Lieutenant T. I. Inglis—had been killed, and 40 non-commissioned officers and men had been wounded.

In their new quarters all ranks of the Battalion now began assiduously to practise the contemplated attack on Wytschaete over ground which had been so prepared as to resemble its main features as far as possible, but this training lasted barely a fortnight, for on the 15th the Brigade was again on the march, and moved by Arques, Hazebrouck, and Bailleul to Doncaster Huts at Locre.   At Bailleul the 48th Brigade marched past Lieutenant-General Sir A. Godley, himself an old Royal Dublin Fusilier, and now commanding the Second Australian and New Zealand Army Corps.   The G.O.C.'s 16th Division and 48th Brigade were also present.   The following message was afterwards published :—" *The General Officer Commanding wishes to express his appreciation of the excellent turn-out and marching of the Brigade, as it marched past Lieutenant-General Sir A. J. Godley and the Divisional Commander to-day.   In view of the persistent bad weather experienced by the troops during the last 14 days, and the difficulties attendant on indifferent and scattered billets, this reflects the greatest credit on all ranks.*"

On the night of May 2nd-3rd the Battalion carried out a small raid for the purpose of obtaining prisoners and gaining identifications.   The raiders were divided into three parties—the first, of 8 men, under Second-Lieutenant Beaumont ; the second, of 12, led by Second-Lieutenant Alexander ; and the third, comprising

1 Lewis gun and detachment and 1 stretcher and bearers, under command of Second-Lieutenant Donovan. Zero hour was at 2.45 a.m., when No. 1 party left the trenches and entered those of the enemy, while No. 2 blocked their own point of entry and moved towards No. 1. The raiders were unfortunately observed by the enemy's sentry post, who sent up lights and opened fire, while an attempt to rush the enemy post was frustrated by the arrival of strong German reinforcements, and the raiders had to retire—not before, however, they had accounted for the enemy post and some six others of the defenders. The Battalion casualties were 1 killed, Second-Lieutenant Alexander and 9 men slightly wounded, and 2 missing.

On the 27th of the same month a more serious raid was undertaken by the Battalion, 12 officers and 300 non-commissioned officers and men being employed, besides 10 sappers who were attached to the party. At 10 p.m. salvos from 40 heavy guns opened on Nancy Support, the final objective, while a barrage was also put down by the guns of the field artillery. "B" Company, under Second-Lieutenant Black, reached the first and second objectives at 10.5, while Captain Patterson with "A" Company was at Nancy Support by 10.20. A platoon of "C" Company under Second-Lieutenant Pedlow protected the right flank, while another platoon from "C" Company under Second-Lieutenant Petit guarded the left. The first and second objectives were gained with practically no opposition and prisoners were taken, but between the second and final objectives resistance was greater and bombing contests ensued. The German trenches were found to be very badly damaged, the dug-outs poor and barely splinter-proof.

The raid was entirely successful, 30 prisoners, including an officer, were brought back, at least 50 Germans were killed, and the raiding party was back at 11.24 p.m. with much valuable information obtained from maps, aeroplane photographs and other documents.

The casualties sustained by the raiding party of the Battalion were 2 men killed, Captain R. M. Patterson, Second-Lieutenants G. V. Poulton, D. C. A. Shepard and G. S. Falkiner and 41 other ranks wounded, while Second-Lieutenants J. P. Scott and A. Holmes and 6 men were missing.

The strength of the 2nd Battalion Dublin Fusiliers at the end of May, just prior to the commencement of the attack on the Messines—Wytschaete Ridge, was 47 officers and 898 non-commissioned officers and men.

" The group of hills known as the Messines—Wytschaete Ridge lies about midway between the towns of Armentières and Ypres. Situated at the eastern end of the range of abrupt, isolated hills which divide the valleys of the River Lys and the River Yser, it links up that range with the line of rising ground which from Wytschaete stretches north-eastwards to the Ypres—Menin road, and then northwards past Passchendaele to Staden. The village of Messines, situated on the southern spur of the ridge, commands a wide view of the valley of the Lys, and enfiladed the British lines to the south. North-west of Messines the village of Wytschaete, situated at the point of the salient and on the highest part of the ridge, from its height of about 260 feet commands even more completely the town of Ypres and the whole of the old British positions in the Ypres salient.

" The German front line skirted the western foot of the ridge in a deep curve from the River Lys opposite Frelinghien to a point just short of the Menin road. The line of trenches then turned north-west past Hooge and Wieltje, following the slight rise known as the Pilckem Ridge to the Yser Canal at Boesinghe. The enemy's second-line system followed the crest of the Messines—Wytschaete Ridge, forming an inner curve."*

The British preparations for the attack were exceptionally elaborate, and had been difficult to carry out since everywhere the German positions overlooked our lines. Roads had been made, railways had been laid down—all across country which had been almost completely destroyed by shell-fire. A great length of pipe line had been constructed for the conveyance of drinking water to the most forward areas, while mining had been carried out on an

* Sir D. Haig's despatch of December 25th, 1917.

G

exceptionally huge scale for the explosion of nineteen great mines at the very moment of the assault. Some of these mines had been completed many long months prior to the attack, and constant and anxious work was necessary to ensure that they remained undetected and undisturbed by the enemy.

In addition to the defences of the ridge itself, two chord positions had been constructed across the base of the salient from south to north. The first lay slightly to the east of the hamlet of Oosttaverne, and was known as the Oosttaverne Line. The second chord position, known as the Warneton Line, crossed the Lys at Warneton, and ran roughly parallel to the Oosttaverne Line a little more than a mile to the east of it.

" The natural advantages of the position were exceptional, and during more than two years of occupation the enemy had devoted the greatest skill and industry to developing them to the utmost. Besides the villages of Messines and Wytschaete, which were organised as main centres of resistance, numerous woods, farms, and hamlets lent themselves to the construction of defensive posts. . . . The actual front selected for attack extended from a point opposite St. Yves to Mount Sorrel inclusive, a distance following the curve of the salient of between 9 and 10 miles. Our final objective was the Oosttaverne Line, which lay between these two points. The greatest depth of our attack was therefore about $2\frac{1}{2}$ miles.

" The British front was here held by the Second Army commanded by General Sir Herbert Plumer, and the portion immediately to be assailed was opposed by three of the six corps of which his army was composed. From opposite Mount Sorrel, astride the Ypres—Comines Canal to the Grand Bois just north of Wytschaete, lay the Xth Corps, under General Morland, with a north-country division on its extreme left, and the 47th London Division in the centre Opposite Wytschaete was the IXth Corps under General Hamilton-Gordon, containing Welsh and West of England troops, and the 16th Irish Division. South lay the Second Anzac Corps, under General Godley, with the Ulster Division on its left, the New

[Photo, Imperial War Museum.

BATTLE OF MESSINES RIDGE. SMASHED GERMAN FORTRESS AND TRENCHES. JUNE 11TH, 1917.

[Photo, Imperial War Museum.

PANORAMIC VIEW OF YPRES, OCTOBER 1ST, 1917.

Zealand Division as its centre, and the 3rd Australian Division on its right astride the Douve. The two southern corps had the task of the direct assault on the ridge, while the Xth Corps, with a much longer front, had to clear the hillocks towards the Ypres salient, and advance upon the ridge and the Oosttaverne line from its northern flank."*

" Messines was distinguishable in the current account as an ' Irish Day.' The Wytschaete end of the ridge was taken by Irish battalions, in which were Ulster men and South of Ireland men, men who take different sides in the religious and political controversies which separate Ireland, but who take a common view as to the necessity of ending the German threat to human civilisation."†

On the evening of Wednesday, June 6th, there was a very heavy thunderstorm and torrents of rain fell, but towards morning this ceased and the day broke tolerably clear on the 7th, though misty in the hollows. The British bombardment had slightly slackened in the night, but the enemy seemed to fear that this might be merely the precursor of the storm, and he sent up signals calling for a barrage and his guns belched forth shrapnel and high explosive.

At 9.30 p.m. on the 6th the 2nd Battalion Royal Dublin Fusiliers left Clare Camp, where since the 2nd it had been engaged in training, and moved to Assembly Trenches, and by 2 a.m. on the 7th had arrived at Vierstraat Switch where the companies were distributed as follows :—" A " Company, Fosse to Watling Street, both inclusive ; " C " and " D " Companies, north of Watling Street, " C " Company being on the right. Battalion headquarters were at the Fosse.

At 3.10 a.m. precisely the nineteen great mines, containing well over one million pounds of high explosive, exploded, and, as stated by the correspondent of the *Morning Post*, " the ground opened below the brim of Wytschaete as though the furnace door of Hades itself had been thrown back, and an enormous fountain of

---

* Buchan, Vol. XX, pp. 60, 61.
† Fox, *The Battles of the Ridges*, p. 103.

fire spurted into the sky. It was infinitely terrifying. Another fountain gushed on the right and still another on the left. The black shadows of the slopes turned to flame as the mines went up, and with them stout trenches. The scarred face of Hill 60, a score of redoubts, and long banks of wire vanished in the night, and when the dawn came there remained only dreadful gaps and fissures in the naked earth, where our men afterwards found embedded the bodies of many of their enemies. The earth rocked as though a giant hand had roughly shaken it. Then the guns began anew," and while the *débris* of the explosion still hung in the air the British divisions of assault went over their parapets.

The Battalion remained in reserve, following up in artillery formation to Chinese Wall, and at 9.15 a.m. occupying Watling Street to Usnagh Trench with headquarters at Shantung. The next two days were spent in consolidating the new line running north from Wytschaete, for, as stated by General Sir Douglas Haig, " in this attack we captured the villages of Messines and Wytschaete and the enemy's defence system, including many strongly organised woods and defended localities on a front of over 9 miles from south of La Douve brook to north of Mount Sorrel. Later in the day our troops again moved forward, in accordance with the plan of operations, and carried the village of Oosttaverne and the enemy's rearward defence system east of the village on a front of over 5 miles."

The 48th Brigade came out of the line on the night of June 9th and took up a position in Vierstraat Switch. During the preceding days the Battalion suffered no losses, but since June 1st it had had 7 men killed, Second-Lieutenants A. J. Clancy and M. Gaffney and 24 non-commissioned officers and men wounded ; while on the 11th at 3 a.m. a 5.9 shell burst in the house occupied as Battalion Headquarters, slightly wounding Lieutenant-Colonel Jeffreys, the commanding officer. On this date the Battalion was taken out of the line for rest and training, proceeding to the Merris area, and being first stationed at Flêtre, then at Westoutre, and later at Sulvestre Cappell ; while here the 16th Division was transferred from the IXth Corps, Second Army, to the XIXth Corps,

Fifth Army, the XIXth Corps being commanded at this time by Lieutenant-General H. E. Watts, C.B., C.M.G., and the Fifth Army by General Sir H. Gough. This transfer necessitated another move, on June 22nd, to the Rubrouck area.

The following telegram and orders followed upon the operations of June 7th—8th: the telegram from His Majesty the King was forwarded by the Commander-in-Chief to General Plumer :—

" I rejoice that thanks to thorough preparation and splendid co-operation of all ranks the important Messines Ridge which has been the scene of so many memorable struggles is again in our hands. Tell General Plumer and the Second Army how proud we are of the achievement by which in a few hours the enemy was driven out of strongly-entrenched positions held by him for two and a half years."

On June 8th Sir Douglas Haig issued the following Order of the Day :—

" The complete success of the attack made yesterday by the Second Army under the command of General Sir Herbert Plumer is an earnest of the eventual final victory of the Allied cause.

" The position assaulted was one of very great natural strength, on the defences of which the enemy had laboured incessantly for nearly 3 years. Its possession overlooking the whole of the Ypres salient was of the greatest tactical and strategical value to the enemy. The excellent observation which he had from this position added enormously to the difficulty of our preparations for the attack, and ensured to him ample warning of our intentions. He was therefore fully prepared for our assault and had brought up reinforcements of men and guns to meet it. He had the further advantage of the experience gained by him from many previous defeats in battles such as the Somme, the Ancre, Arras and Vimy Ridge. On the lessons to be drawn from these he had issued carefully thought-out instructions. Despite all these advantages the enemy has been completely defeated. Within the space of a few hours all our objectives were gained, with undoubtedly very severe loss to the Germans. Our own casualties, were, for a battle of such magnitude, most gratifyingly light.

" The full effect of this victory cannot be estimated yet, but that it will be very great is certain.

" Following on the great successes already gained it affords final and conclusive proof that neither strength of position nor knowledge of and timely preparation to meet impending assault can save the enemy from complete defeat, and that brave and tenacious as the German troops are, it is only a question of how much longer they can endure the repetition of such blows. Yesterday's victory was due to causes which always have given and always will give success, viz. : the utmost skill, valour and determination in the execution of the attack following on the greatest forethought and thoroughness in preparation for it.

" I desire to place on record here my deep appreciation of the splendid work done, above and below ground as well as in the air, by all Arms, Services, and Departments, and by the Commanders and Staffs by whom, under Sir Herbert Plumer's orders, all means at our disposal were combined, both in preparation and in execution, with a skill, devotion and bravery beyond all praise.

" The great success gained has brought us a step nearer to the final victorious end of the war, and the Empire will be justly proud of the troops who have added such fresh lustre to its arms."

Major-General Hickie, commanding the 16th Irish Division, published the following :—

" The 16th Division has been relieved to-day. For over 8 months we have held our portion of the line in the Spanbrock, Vierstraat or Diependaal sectors, and we can look back with pride and satisfaction to the record of those months. Neither rain, nor snow or the heat of summer has interfered with the constant work. Long distances, wet roads, cold nights, shortage of fuel, hostile shelling, have all failed to damp the spirits of our men. With gun and howitzer, trench mortar, rifle, machine gun and bomb, and sometimes with bayonet, we have gradually worn down the boasting enemy ; and 2 days ago, with small losses to the Division, we have completed this chapter of our history by taking from him the Wood and Village of Wytschaete and the crest of the hill which means so much for future operations.

" The Divisional Commander, in congratulating all arms and all ranks of the Division on their victory of June 7th, thanks all the Officers, Non-Commissioned Officers and Men under his command for their loyalty and help and for their bravery and skill in action.

" Whatever new work lies before us, it will be tackled with the same endurance, the same cheerfulness and the same bravery ; and again and again the Division and every man in it will justify the right to our motto—' Everywhere and always faithful.' "

During the month of July His Majesty the King paid a visit to his army in France and Flanders and at the conclusion of his tour the following gracious message was promulgated to all ranks :—

" Officers, Non-Commissioned Officers and Men.

" On the conclusion of my fourth visit to the British Armies in the field, I leave you with feelings of admiration and gratitude for past achievements and of confidence in further efforts.   On all sides I have witnessed the scenes of your triumphs.   The battlefields of the Somme, the Ancre, Arras, Vimy and Messines have shown me what great results can be attained by the courage and devotion of all arms and services under efficient commanders and staffs.

" Nor do I forget the valuable work done by the various departments behind the fighting line, including those who direct and man the highly-developed system of railways and other means of communication.

" Your comrades, too—the men and women of the industrial army at home—have claims on your remembrance for their untiring service in helping you to meet the enemy on terms which are not merely equal but daily improving.

" It was a great pleasure to the Queen to accompany me, and to become personally acquainted with the excellent arrangements for the care of the sick and wounded, whose welfare is ever close to her heart.

" For the past three years the Armies of the Empire and workers in the Homelands behind them have risen superior to every difficulty and every trial.   The splendid successes already gained, in concert with our gallant Allies, have advanced us well on the way towards the completion of the task we undertook.   There are doubtless fierce struggles

still to come and heavy strains on our endurance to be borne. But be the road before us long or short, the spirit and pluck which have brought you so far will never fail, and, under God's guidance, the final and complete victory of our just cause is assured."

During the month of July there was one casualty only, a man of the Battalion being killed in action on the 31st, while there came out in drafts of varying strengths 2 officers—Captain Trigona and Second-Lieutenant Humphrey—and 109 non-commissioned officers and men. On the 21st of this month the effective strength of the Battalion was 42 officers and 850 other ranks ; but since the actual fixed war establishment was 30 officers and 950 non-commissioned officers and men, it is apparent that the Battalion was on this date 12 officers over, and 114 other ranks below establishment.

The Battalion did not return to an active area until the end of July, when it was in occupation of St. Lawrence Camp just west of Ypres.

The operations, which at that period of the war and for some time after were known as the Third Battle of Ypres and which had their commencement on July 31st, were part of the general plan of a Flanders offensive which Sir Douglas Haig had evolved the previous winter, and which included an offensive east and north of Ypres. The attack now projected involved a complete redistribution of the Allied forces. " The front of the Third Army was greatly extended, and now covered all the ground between Arras and the junction with the French. This released Sir Hubert Gough's Fifth Army and Sir Henry Rawlinson's Fourth Army for service in the north. In early June French troops had held the front on the Yser between St. George's and the sea ; these were now relieved by the British Fourth Army. . . . From Boesinghe to the Zillebeke—Zandvoorde road south-east of Ypres lay the British Fifth Army, and on its right the Second Army as far as the Lys. From Armentières to Arras Sir Henry Horne's First Army held the front."*

* Buchan, Vol. XX, pp. 76, 77.

[*Photo, Imperial War Museum.*

A SCENE ON THE BATTLEFIELD BEYOND YPRES, ST. JULIEN.

[*Photo, Imperial War Museum*

A LINE OF GERMAN STRONG POSTS, NEAR ST JULIEN.

"So effective was our counter-battery work that the enemy commenced to withdraw his guns to places of greater security. On this account . . . the date of our attack, which had been fixed for 25th July, was postponed for 3 days. . . . Subsequently a succession of days of bad visibility, combined with the difficulties experienced by our Allies in getting their guns into position in their new area, decided me to sanction a further postponement until 31st July."*

On July 30th the 2nd Battalion Royal Dublin Fusiliers was in the position of readiness at St. Lawrence Camp, and moved at 3.35 a.m. on the following day to the Assembly Trenches, finding itself on August 1st in Liverpool Trench under the orders of the Brigadier-General Commanding the 164th Brigade. The following parties were detailed from the Battalion :—From "C" Company, 1 officer and 25 other ranks to carry up trench mortar ammunition ; and from "B" Company, 1 officer and 50 non-commissioned officers and men to act as stretcher-bearers.

At 8 p.m. that evening the Battalion was relieved and moved back again to St. Lawrence Camp, but on the following day was sent up to the front-line trenches, the Battalion headquarters being at Frezenberg. On the way up the companies had to pass through a very heavy enemy barrage, and the men were obliged to shelter in shell-holes for upwards of an hour and a half. On reaching the front-line trenches, "A" and "B" Companies were in front, "C" in support, and "D" in reserve, but since the whole Battalion occupied concrete dug-outs the cover obtained from enemy fire was tolerably complete. Here the companies remained until the 7th, when the Battalion was withdrawn to camp in the Vlamertinghe area ; but about midnight, while moving back over the old No Man's Land, the enemy put over very many mustard-gas shells, causing several casualties among all ranks, Major Smithwick being admitted to hospital from gas poisoning, and Captain Duff-Taylor assuming command of the Battalion in his place. The Battalion remained in this area until the morning of August 15th, on which date the

* Field-Marshal Haig's despatch of December 25th, 1917.

48th Infantry Brigade was thus disposed in the right sector of the 16th Division front :—

> Front Line : 2nd Battalion Royal Dublin Fusiliers, head-quarters at Frezenberg Redoubt.
>
> Support Line : 1 company 8th Battalion Royal Dublin Fusiliers, headquarters at Wilde Wood.
>
> Vlamertinghe : 7th Battalion Royal Irish Fusiliers and 9th Battalion Royal Dublin Fusiliers, being the two battalions detailed to carry out the assault.

During the past few days, however, the casualties sustained by the Brigade were many from shell-fire, gas poisoning, and sickness—the latter due to very adverse weather conditions, rain having fallen continuously during four days and nights, and the strength of the Brigade was reduced by 27 officers and nearly 700 other ranks. Consequently " D " Company (Captain Addis) 2nd Royal Dublin Fusiliers had been broken up, and the officers and men attached to " A," " B," and " C," while the 8th Battalion Royal Dublin Fusiliers, after finding all carrying parties, could only muster one company, and this, under Major Cowley, was attached for the operations to the 2nd Battalion.

A battalion—the 1st Royal Munster Fusiliers—was attached to the 48th from the 47th Infantry Brigade.

The strength on this day of the Battalion was 14 officers and 293 other ranks ; that of the company of the 8th was 6 officers and 85 other ranks. Second-Lieutenant Stone, of the Battalion, had just been detailed for duty with the 48th Trench Mortar Battery, and was in charge of the section attached.

The Battalion was thus disposed :—" A " and " B " Companies, under Captains Black and Byrne respectively, held the right and left of the front line ; " C " Company (Captain Burrowes) occupied a trench in the vicinity of Frezenberg Redoubt ; while Major Cowley's company of the 8th Dublin Fusiliers was in Douglas Trench.

On the evening of the 15th the two assaulting battalions proceeded to the " jumping-off " positions assigned to them, but the

R.E. officer, who was to have guided them and taped out the front line of each wave, was wounded and his men scattered, so they were eventually led to their places by guides supplied by Captains Black and Byrne, with the result that there was none of the confusion and delay that might reasonably have been expected, and the two battalions were in position on schedule time—viz., 4 a.m. on the 16th—despite heavy shelling throughout the night all along the Frezenberg Ridge.

At zero hour " A " Company of the Battalion was in rear of the leading wave of the Royal Irish Fusiliers, while the remaining two companies occupied the positions already stated above.

Some little time previous to zero hour the enemy had opened a heavy barrage whereby many casualties were incurred, and during their advance to the assault the Royal Irish Fusiliers and 9th Battalion Dublins sustained losses amounting to 64 and 66 per cent. respectively of their assaulting strength. The two companies of the 2nd Battalion detailed to " mop up " advanced in rear of the assaulting battalions to Vampire Farm and Potsdam, and other trenches in the vicinity of these two points, but both suffered severely from machine-gun and shell-fire before reaching their objectives. The company on the left—" B," under Captain Byrne— was practically annihilated before reaching Vampire Farm, only two officers and three other ranks surviving, and a message coming back to headquarters that Vampire Farm contained five machine guns. " A " Company (Captain Black) succeeded in reaching the support company of the Royal Irish Fusiliers, which was found to be held up by enemy fire, and by this time all the officers with " A " Company were casualties.

Those who were left of these battalions dug in on the line they had reached, but at 4.10 p.m., owing to the withdrawal of a battalion of the Brigade on the left, the remainder of " A " Company, now reduced to 1 non-commissioned officer and 6 men, was obliged to fall back to the original line.

At 9.30 a.m. " C " Company of the Battalion, in accordance with orders from the Brigade, moved up in support of the 9th

Dublin Fusiliers, which had suffered severely from intense machine-gun fire, and succeeded in getting to within 100 yards of Bit Work, when only 2 officers and 10 men remained. At 10.30 Major Cowley moved forward two platoons in support of each assaulting battalion, and though these lost all their officers and incurred heavy losses, they held on to the position reached until 10 p.m. on the 17th, when, unable to consolidate, unsupported on either flank, and raked by machine-gun fire in front and in enfilade, they also at last withdrew.

On the 17th the Brigade was relieved and moved back to camp in the Vlamertinghe area, having lost between August 1st and 17th 82 officers and 1,550 other ranks.

In concluding his report on these operations, the Brigadier wrote : " I can only place upon record my great appreciation of the spirit evinced by all ranks during the occupation of the line for 12 continuous days prior to the attack, during the action and subsequently. The Brigade failed to maintain the objectives reached, but I think that the percentage of casualties which I have noted against units, and the fact that the flanks were exposed, will suffice to explain why."

The 2nd Battalion Royal Dublin Fusiliers lost in the operations of the 16th and 17th 50 per cent. of the officers and men who went into action. The actual casualties from the 1st to 17th were :—
Killed—Second-Lieutenants G. S. Falkiner, A. W. Storrar, W. L. Reavie, J. Stewart, G. L. Graham, and 26 other ranks ; *wounded*—Lieutenant H. G. Aylmer, Second-Lieutenants W. I. Black, A. E. Burrowes, and 127 non-commissioned officers and men ; *gassed*—Major S. G. Smithwick, Second-Lieutenants F. Fennell, W. Pedlow, W. Harney, and 173 other ranks ; *missing* — 23 men ; and *shell-shocked*—2.

The Brigade had presumably been withdrawn with the view of affording its units some sorely-needed rest, but they were much moved about during the few days they remained out of the line. Thus on the 18th the Battalion went by train to Poperinghe and camped in the Watou area ; on the 20th it marched to billets at

[Photo, Imperial War Museum.

HOTEL DE VILLE, ARRAS.

THE BATTLE OF FLANDERS. SCENE ON THE BATTLEFIELD, ST. JEAN.
JULY 31st, 1917.

Wormhoudt ; on the 21st it was at Bapaume, moving next day to Courcelles-le-Comte to reorganize.   On the 26th the Battalion found itself at Moyenneville, and was on the following day in support of the VIth Corps centre sector, when the companies were disposed as follows :—

  " A " Company : At the quarry.
  " B " Company : In the road at T.23.A.
  " C " Company : In posts C 6-10 inclusive.
  " D " Company : In Shaft Trench.
  Headquarters : At T.22.d.69.

Here the Battalion remained quietly until the end of August, when it was withdrawn and sent until September 16th to billets at Ervillers.   From the 17th to the 22nd the 2nd Royal Dublin Fusiliers were again in the front line, and during that time the German guns were especially active, on one day some 200 yards of trench held by " D " Company being totally obliterated.

There was nothing of very particular note that happened during the early part of October, but on the 21st a very dashing and successful raid was carried out by a party of the Regiment against the enemy's works at the apex of Prince and Tunnel Trenches with the object of obtaining identifications, etc.   The party consisted of Lieutenant Beaumont and Second-Lieutenant Lloyd, with 15 non-commissioned officers and men.

The raid was very successful, and there was only one casualty in the party, one man being slightly wounded.

Congratulations came to the Battalion from all sides.

From the General :

" *Please convey my congratulations to Colonel Jeffreys and the officers and men who took part in to-day's excellently planned and well carried out raid.*"

From the General Officer Commanding the VIth Army Corps :

" *This small raid was well planned and carried out, and is all the more satisfactory as it was entirely initiated by the Battalion Commander.*"

While the following message was received from the G.O.C.
Third Army :

*" A very dashing and successful exploit.   The 2nd Royal Dublin
Fusiliers are to be congratulated "*

Theoretically the so-called Third Battle of Ypres came to an
end in the first week in November, but there was further fighting
before the year 1917 finally closed.   Thus on November 17th the
Battalion again found itself occupying preliminary assembly positions preparatory to an attack upon the enemy front, being disposed
as follows :—

"A" and "B" Companies : In the Front Line.
"C" and "D" Companies : In Lincoln Support.

But by 5.30 a.m. on the 20th all four companies were in the front—
"A" and "C," under Captains Pedlow and Jameson, being on the
right ; with "B" and "D," under Captains Byrne and Addis, on
the left.   Each company had detailed a platoon to remain in support, and these were assembled in Lincoln Support under Captain
Delany.

The attack commenced at 6.30 a.m., when the enemy was
completely taken by surprise, and not very much resistance was
offered except by those in charge of the machine guns in Mebus
Juno, where the issue was fought out to the bitter end.   Tunnel
Trench was found to have been mined, but the mines were cut,
and the German officer in charge was captured by the men of
"A" and "C" Companies of the Battalion.   During the rest of
the day there were repeated threats of counter-attacks, but these
did not develop, and the ground gained was consolidated under the
direction of Major Stirke.

The casualties during this action were by no means light,
Second-Lieutenant J. A. Harvey and 9 men being killed, while
Second-Lieutenant J. J. Perry and 54 other ranks were wounded
and there were 3 missing.

Having been relieved in the front line on November 22nd, the
Battalion went back to Belfast Camp, Ervillers, moving on

December 1st south to the Achiet-le-Petit area (with the 16th Division, owing to the success of the German counter-attack after Cambrai), and thence again on the 4th to St. Emilie, where the companies took over the Lempire defences. There were constant reports of coming attacks, but these did not materialize.

During the month of December the casualties in the Battalion totalled 4 killed, 15 wounded, and 3 missing, while against this wastage the drafts amounted only to 2 officers and 45 non-commissioned officers and men.

At the end of the year the Battalion was in divisional reserve at Trincourt.

During the last few weeks the 48th Infantry Brigade had been almost entirely reconstituted, consequent on the very heavy losses experienced by its units, and, indeed, by those of the 16th Division generally, in the severe fighting of the previous August, and by the fact that the Division found itself unable to maintain its strength and its special character without absorbing some of the Irish battalions in other divisions. At the end of October the 1st Battalion Royal Dublin Fusiliers had joined the 48th Brigade from the 86th Brigade of the 29th Division, and the 48th Brigade was now composed as under :—

> 1st Battalion Royal Dublin Fusiliers.
> 2nd Battalion Royal Dublin Fusiliers.
> 7th Battalion Royal Irish Fusiliers.
> 8th Battalion Royal Dublin Fusiliers.
> 9th Battalion Royal Dublin Fusiliers.

CHAPTER VI

1918.

THE BATTLES OF THE SOMME AND OF LE CATEAU.

THE END OF THE WAR.

DURING the months of January and February, 1918, the 2nd Battalion Royal Dublin Fusiliers remained in and about the St. Emilie defences, but already as early as the middle of February there were persistent rumours that the enemy was elaborating measures for an attack upon an unusually large scale. It is true that in nearly every one of the more distant theatres of war the Allies had secured many and substantial successes, but these distant campaigns had only a remote and indirect effect upon the progress of the war in Europe, and here the balance of power had tilted dangerously against the British and their comrades in arms. Russia had completely broken down, the country had been over-run by the Germans, and Roumania, with enemies in front and traitorous friends on her rear and flank, had been compelled to make terms. As a result of these defections the whole strength of the German and Austrian forces was available for the war in the west, and during the winter months of 1917–18 troop trains were hurrying German divisions from the Russian frontiers to the plains of France and Flanders. In this extremity many of the British and French divisions which had been sent to stiffen the Italian line after the disaster of Caporetto, had been hastily recalled, despite the fact that a great Austrian army was confronting the Italians on the Piave.

Such patrols as were sent out from the 16th Division failed, however, to detect signs of any serious enemy reinforcement in the immediate front.

For some little time past the heavy and increasing drain on

the manhood of the nation had been causing the military authorities very considerable anxiety, and it was finally reluctantly determined to carry out a reorganization of the divisional composition of the British Army in France ; this measure was completed during the month of February, 1918, when the number of battalions in a division was reduced from thirteen to ten, and the number of battalions in each brigade from four to three. The effect of this reorganization upon the 16th Irish Division was that the 8th and 9th Battalions of the Royal Dublin Fusiliers were practically disbanded and taken out of the Division, and that the 48th Brigade was now composed of the 2nd Battalion Royal Munster Fusiliers— transferred from the 1st Division—and the 1st and 2nd Battalions of the Royal Dublin Fusiliers.

Early in March the enemy's artillery and raiding parties displayed unusual activity, while on our side many counter-raids were carried out with the object of obtaining all possible identifications of the new German divisions said to be moving up.

On the 16th of this month the 2nd Battalion was thus disposed :
  " A " Company in Yak and Zebra Posts and Rose Trench.
  " B " Company in Ridge Reserve Support.
  " C " Company in Sandbag Alley and part of Ridge Reserve
        South.
  " D " Company in Enfer Wood and Quid Trench.

From these positions a very successful raid was conducted on the night of March 19th—20th by a party of 30 men of the Battalion led by Lieutenant Addis and Second-Lieutenant Quigley ; these entered the German trenches north of Lark Post, killed six Germans, wounded several others, and retired bringing with them a wounded prisoner. Two men of the Dublin Fusiliers party were slightly wounded.

It was by this time established beyond all possible doubt that the Germans were preparing for an offensive on a very large scale, and that their main attack was intended to fall upon that part of the British line held by the Third and Fifth Armies, each of which consisted of approximately fifteen divisions. The front occupied

H

by these Armies, commanded respectively by Generals Sir Julian Byng and Sir Herbert Gough, was a long curve of some 50 miles from Arras in the north down to Barisis, 8 miles south of La Fère, and was strongly fortified throughout its length.

" The Third Army consisted of four corps, the XVIIth (Fergusson) in the Arras—Monchy sector, the VIth (Haldane), carrying the line past Bullecourt, the IVth (Harper), continuing it to near the Cambrai district, and the Vth (Fanshawe) covering that important point where the gap in the Hindenburg line seemed to make an attack particularly likely. The Fifth Army in turn consisted of the VIIth Corps (Congreve) in the southern part of the Cambrai district, the XIXth (Watts) from south of Ronssoy to Maissemy, the XVIIIth (Maxse) in front of St. Quentin, and the IIIrd Corps (Butler) covering the great frontage of 30,000 yards from Urvillers, across the Oise, down to Barisis."*

" The general principle of our defensive arrangements on the front of those armies," wrote the Field-Marshal in his despatch of October 21st, 1918, " was the distribution of our troops in depth. With this object three defensive belts, sited at considerable distances from each other, had been constructed or were approaching completion in the forward area, the most advanced of which was in the nature of a lightly held outpost screen covering our main positions. On the morning of the attack the troops detailed to man these various defences were all in position. Behind the forward defences of the Fifth Army, and in view of the smaller resources which could be placed at the disposal of that Army, arrangements had been made for the construction of a strong and carefully-sited bridge-head position covering Peronne and the crossings of the River Somme south of that town. Considerable progress had been made in the laying out of this position, though at the outbreak of the enemy's offensive its defences were incomplete. . .

" In all at least 64 German divisions took part in the operations of the first day of the battle, a number considerably exceeding the total forces composing the entire British Army in France. . . . To

* Conan Doyle, Vol. V, p. 10.

THE STRAND, PLOEGSTEERT WOOD.

A SCENE ON NEWLY CAPTURED GROUND.

AMMUNITION WAGGON GOING UP WHILE GUNNERS ARE GETTING A BIG GUN
INTO A NEW ADVANCE POSITION FOR BOMBARDING THE BOCHE, NEAR ST.
JULIEN, OCTOBER 17TH, 1917.

meet this assault the Third Army disposed of eight divisions in line on the front of the enemy's initial attack, with seven divisions available in reserve. The Fifth Army disposed of fourteen divisions and three cavalry divisions, of which three infantry divisions and three cavalry divisions were in reserve. The total British force on the original battle front, on the morning of March 21st, was twenty-nine infantry divisions and three cavalry divisions, of which nineteen infantry divisions were in line."

At this time the front of the VIIth Corps—14,000 yards—" went southward from a point about half-a-mile north of Gouzeaucourt, at the top of a hill about 400 yards west of Gonnelieu, through Gauche Wood to Vaucellette Farm, then south-eastward to Epéhy and Ronssoy."*

On March 21st when the great German attack opened the 16th Division had two brigades in the line, the 48th on the left, and the 49th on the right, and these sustained an assault of a peculiarly crushing character.

At 4.45 a.m. on the 21st a violent bombardment was opened by the enemy from all his guns ; this continued until 11.15 when he attacked along the whole front. By this time the battalions had all assumed their battle positions, the 2nd Royal Dublin Fusiliers being on the right, the 1st Battalion in the centre, and the 2nd Battalion Royal Munster Fusiliers on the left, each battalion holding the front with two companies in " the Outpost Zone " and the remaining two companies in " the Main Battle Position." The front line was in good order and well wired, while in rear of this was a series of strong points—roughly three of them on each battalion front—at from 200 to 600 yards in rear of the front line. These strong points were completely wired in and were garrisoned by one or two platoons according to the size and tactical importance of the post.

The Main Line of Resistance, known as " the Red Line," ran along the ridge between Epéhy and Ronssoy and in front of the latter village, and varied in distance from the front line from 1,200

* Sparrow, *The Fifth Army in March*, 1918, p. 19.

H 2

to 2,500 yards ; this system was composed of a front and support line connected by communication trenches. Dug-outs were few in number. In rear of this again were " the Yellow Line " running in front of Ronssoy Wood, " the Brown Line " in front of St. Emilie, and " the Green Line " in front of Tincourt and Hamel. These had all been wired, but only the first two had actually been dug, the last of the three having merely been " spitlocked " or marked out.

The Battalion had already been some 40 days in the trenches and was overdue for relief when the German attack became imminent, and relief was then postponed *sine die*.

The following was the disposition of the companies of the Battalion on the early morning of March 21st, the day the German attack opened :—

" *A* " *Company* : Captain Cunningham ; 1 platoon Yak Post (Second-Lieutenant Jackson) ; 1 platoon Zebra Post (Second-Lieutenant O'Connell) ; 2 platoons and Company Headquarters Rose Trench (Lieutenant Addis).

" *C* " *Company* : Captain Karney, Second-Lieutenants Petit and Wilkin ; 3 platoons at Ridge Reserve ; 1 platoon and Company Headquarters in Ronssoy.

" *D* " *Company* : Captain Addis ; 1 platoon in Enfer Wood— an isolated post in front of "the Red Line " ; 2 platoons in Ridge Reserve ; 1 platoon in Quid Post.

" *B* " *Company* : Captain Lawrence, Lieutenant Lloyd and Second-Lieutenant Quigley, in support in Ridge Support.

Battalion headquarters was in Ronssoy when the enemy launched his attack, but then moved up to Ridge Support in accordance with the defence scheme.

On the morning of the 21st the effective strength of the Battalion was 23 officers and 643 non-commissioned officers and men ; Major Wheeler was in command, having taken over from Major Stirke, Captain Byrne was second in command, Captain Stitt was adjutant, Major Burke quartermaster, while the Rev. Father Casey was chaplain.

The bombardment which, as already stated, opened from the German lines at 4.45 a.m. was of the nature of a barrage extending to a depth of 8 miles ; it was very dark and a thick fog prevailed ; within five minutes of the opening of the bombardment all telephonic communication was cut, visual communication was out of the question by reason of the fog, and such was the intensity of the fire that no runners could get through. Indeed, no movement at all was possible while the shelling lasted, but so soon as this ceased and the mist began to clear about 9 a.m. endeavours were made, but fruitlessly, to get into touch with the forward companies, and it was now discovered that the enemy infantry had almost reached Rose Trench and Ridge Reserve north of Ronssoy ; heavy fighting ensued and the enemy was ejected, having suffered many casualties from the Lewis guns and rifle fire of the Battalion.

About 10 o'clock the Germans, having apparently been re-inforced, attacked again in strength, and succeeded in gaining a footing in Ridge Reserve, north of Sandbag Alley ; here a counter-attack was delivered, the enemy was held up and severely punished, and Second-Lieutenant Quigley and 20 men of the Battalion pushed forward to deal with an enemy machine gun which was giving trouble ; this was destroyed, but Second-Lieutenant Quigley was killed when withdrawing.

During the next two hours very hard fighting took place and great bravery was displayed by those engaged, especially by the bombing parties which pushed along Ridge Reserve to drive out the enemy, and by a few men of " C " Company who were cut off by the entry of the Germans into Ronssoy from the south.

Shortly after this the Ronssoy—Epéhy road, running between Ridge Reserve and Ridge Support, was swept by heavy machine-gun fire, the enemy having broken through the front of the division on the right ; later again machine-gun fire was noticed from the right rear, and a runner was sent to the machine gunners believed to be there, to inform them that the 2nd Dublins were still holding their positions—but it appeared that the fire was that of enemy machine guns.

About 11.45 troops were seen withdrawing through Ronssoy west towards St. Emilie, and a quarter of an hour later numbers of the enemy were noticed in the western outskirts of Ronssoy; as the right flank was thus exposed those troops still holding the right in Ridge Support were ordered to face south and form a flank defence.

For some little time past the German gun-fire had slackened, but it now broke out again with redoubled fury, putting down a barrage of " heavy stuff " along the valley in front of Ronssoy Wood and about 400 yards in rear of and parallel to the position of the Brigade. Matters now began to look very serious; the enemy showed in considerable numbers on the western outskirts of Ronssoy and Ronssoy Wood, thus overlooking the position of the Battalion, and it was evident that the Germans were receiving substantial reinforcement by the east of Ronssoy over ground not covered by British machine-gun and rifle fire.

About 12.10 p.m. a company of the battalion on the left had been forced back and the enemy had worked up to within 600 yards of the Battalion Headquarters and Reserve Company, and was already moving on the right rear of the 2nd Royal Dublin Fusiliers. Satisfied that the position had now become quite untenable, Major Wheeler ordered a withdrawal towards St. Emilie and this was carried out without serious loss covered by the fire of two Lewis guns. The Germans made many attempts to come to close quarters, but were held to their ground by well-aimed rifle fire and suffered many casualties. The retirement was admirably conducted and special praise is due to the Lewis gunners for the steady way in which they took up one position after another and opened covering fire.

On reaching " the Brown Line " the Battalion could muster no more than 7 officers and 200 non-commissioned officers and men.

Shortly after 3 p.m. the artillery of the attackers opened on " the Brown Line " from their heavy guns and this fire was maintained until after dusk. Many stragglers of the Battalion and other corps succeeded in making their way to " the Brown Line " where they were collected and organized in platoons in readiness to

VIEW OF CROISILLES.

SHELL BURSTING IN SQUARE OF A RUINED TOWN IN BELGIUM.

repel an attack which had been noticed to be impending and which was finally delivered about six in the evening ; it was, however, defeated with machine-gun and rifle fire. But elsewhere—as towards Roisel, the enemy was seen to have made some advance.

As darkness began to fall the 2nd Battalion Royal Dublin Fusiliers was relieved on " the Brown Line " by the 6th Connaught Rangers of the 47th Brigade, and then withdrew to the Railway Cutting west of St. Emilie, arriving there very early on the morning of March 22nd.

Very shortly after arrival here—in fact, at dawn—a message was received directing what was left of the Battalion to fall back to the old Divisional Headquarters at Villersfaucon, and on arrival here fresh orders were handed to the commanding officer telling him to move on to Tincourt.   Here rations were issued and the men had their first proper meal since the 20th.   On reorganizing the companies it was found that there remained only 5 officers and 90 other ranks, but during the three following days men joined who had become scattered during the retreat and had attached themselves to other units.

On the night of the 22nd—23rd the Battalion, with the Royal Munster Fusiliers, was formed along the road running north-west from Tincourt supporting a line held by two other infantry battalions and a Field Company, R.E.

Early on the morning of the 23rd an order was received for a general retirement through Courcelles to a position at the Bois des Fleurs covering Bussu and Peronne.   The supporting troops, in which the Battalion was included, held on until the parties from " the Green Line " had passed through them, finally themselves beginning to fall back about 7 a.m.   There was an exceptionally thick mist and it was impossible to see anything at a distance of more than 50 yards at most, so this withdrawal was conducted by compass, and was carried out without molestation, assisted by tanks. It was afterwards stated that this retirement was due to the enemy having broken through on a corps frontage south of the Somme.

The new position east of Doingt was held and an attack made

upon it was beaten off, but the retreat was then resumed to a line west of Peronne which was reached about 3 p.m. Here it was discovered that all troops on either flank of the Brigade had been driven over the Somme, and orders were given to hold the Bridgehead as long as possible, and in the event of being driven back, to take up a position on the hill above Biaches ; finally at 10.30 p.m. fresh instructions were received for the 16th Division to march to Cappy. Major Burke, the quartermaster of the Battalion, with happy foresight, had arranged for the issue of tea to the men on the march and had made arrangements for the provision of a hot meal on arrival at Cappy, which was reached at 4 a.m. on March 24th.

Unfortunately, a platoon of the Battalion left to cover the retirement on the far side of the bridge at Peronne was missing with platoons of other corps.

At Cappy the Battalion was reorganized ; a few reinforcements had come up, and these, with all available men from the Transport, made up four very weak companies, amounting in all to about 7 officers and 120 other ranks. The troops were permitted to get some rest here until about five o'clock in the afternoon, when they were ordered to march to Froissy, crossing the canal south of Bray and holding the bridgehead on the southern bank. Here the Battalion was posted on the left, with the 2nd Battalion Royal Munster Fusiliers on the right and the 1st Dublins in support.

The night passed tolerably uneventfully, and on the 25th the Division was transferred to the XIXth Corps. This transfer was consequent on the decision which had been arrived at to hand over certain divisions to General Byng's Third Army, it being considered that this arrangement would allow General Gough, with the remainder of the Fifth Army, to devote his whole attention to the enemy advancing south of the Somme. The 16th and 39th Divisions, both greatly cut up and exhausted, remained with the Fifth Army, and were included in the XIXth Corps, which, though actually containing six divisions, had no more than the normal strength of two.

" The loss of the line of the Somme was a very serious matter, for the Germans now entered upon the zone in which were placed our depots, stores, and hospitals. These had all to be abandoned or evacuated hastily, and consequently great quantities of war material of all kinds fell into the enemy's hands ; much suffering was caused to the sick and wounded, of whom numbers had to be left untended and without shelter alongside the railway lines in the rear until the hospital trains could pick them up ; the telegraph and telephone communications were disorganized, and the difficulties of organizing defence increased as the danger grew. It was clear that the main object of the Germans was to reach Amiens, and that the weight of their attack was falling upon the Fifth Army."*

In consequence of the new disposal of the Division, the brigades composing it moved forward again and took up a position east of Cappy, covering the bridgehead over the Somme, being ordered, in the event of being driven back, to withdraw towards Proyart. During this night, which passed quietly save for some intermittent shelling, the Battalion occupied a support position facing north.

At daylight on March 26th the troops in front began to retire, and it appeared that the enemy was attacking in strength, and all began to fall back towards Chuignolles, where the Battalion commenced to dig in on a position running from the cemetery south of Chuignolles towards Proyart, with the two battalions of the Royal Munster Fusiliers on the right and left, and the 1st Dublins some 1,500 yards in rear in support. The Battalion had barely dug itself under cover when the Germans attacked, well supported by their heavy artillery ; there was severe fighting all the afternoon, and about 5 p.m., the left flank being turned, the Battalion fell back to a position between Morcourt and Chuignolles, occupying an old French system of trenches. Here the 47th Brigade was on the right of the Battalion, but touch could not be established ; the position was not ideal, being on the forward slope of a hill facing east, with a thick copse on the right.

At daybreak on the 27th the enemy began to shell the position,

* Maurice, *The Last Four Months*, p. 42.

and during the morning it was seen that he was being greatly reinforced, but he was held off by rifle and machine-gun fire, and three guns he had brought into line in an exposed spot were put out of action. Still it was clear that the Germans were making ground on the right flank, which seemed to be completely in the air. The orders received by the Battalion Commander—Major Wheeler—were to hold on and fight to the last. As the day wore on to afternoon, it became increasingly evident that the Battalion and the Munsters on the left were being gradually surrounded, that the resistance made had fulfilled its object, and that no useful purpose would be served by remaining longer in the position. Accordingly Majors Wheeler and Rye (Munsters) consulted together and decided to withdraw at dusk, making for the bridge at Ecluse, which seemed to be the point where other troops would most probably be met with.

The men filed out of the trenches in the dark, and, passing through Morcourt, arrived at the bridge at Ecluse about 8 p.m., and here the movement of transport and sounds of shouting became audible. Captain and Adjutant Stitt, who speaks German, was sent forward with two men to reconnoitre, and returned in a few minutes reporting that the bridge was occupied by the enemy.

The troops now silently turned in their tracks and marched back through Morcourt by the south bank of the canal to the bridgehead at Cérisy, the Dublins leading and the Munsters bringing up the rear. Morcourt, when passed through, was in flames, but, moving on without interference, the vicinity of the bridge at Cérisy was reached, when again troops were seen moving about, and on Captain Stitt going forward once more, the suspicion that those here also were Germans was confirmed. A hurried consultation was held, and the decision was arrived at to rush the bridge, and for this purpose the two battalions were formed up in fours in column of route.

At this moment " it came into my mind," writes Major Wheeler, " that my servant, Private Byrne, was a heavy-weight boxer and had recently won a divisional contest. I called him up and had him placed in the leading section of fours *without his rifle*! He knew without being told what he was likely to have to do. When all was ready we moved off, Captain Stitt and I leading. The Germans showed no sign of suspecting anything until we were within some 15 yards of them, when they challenged. We immediately rushed forward, Fusilier Byrne knocking out two Boches with a right and left ;* a couple of revolver shots accounted for two more, and so completely had we taken this picquet by surprise that the remaining Germans made themselves scarce.

" Having cleared the Bridgehead and got the men again into fours, we continued our march without interference, but from now on we had a very hazy idea of our direction, and it was more by good luck than good management that all turned out well ; one of our heavy guns was dropping shells round Cérisy, and we marched more or less towards the position indicated by the report of the firing.

" We next found ourselves at a village which turned out to be Sailly-Lorette, where we again bumped into a German picquet, and as they were taken completely by surprise they had small chance of defending themselves, and were all killed in a hand-to-hand fight. Not knowing how many Germans were about, and the probability being that the village was full of them, we quickly disappeared."

The march was now continued unmolested, and at 2 a.m. on the 28th the little column arrived at the Bois de Hamel and went to billets in the village, having marched over 12 miles through the German lines.

Later in the morning Brigadier-General Ramsay ordered a special parade, and " warmly congratulated the Battalion on the brilliant stand it had made on the 27th, and the excellent manner in which it had evaded the clutches of the enemy, bringing with it the remnants of the other units—in all, about 250 officers, non-commissioned officers and men."

The fighting strength of the Battalion was by this reduced to 3 officers—Major Wheeler, Captain Stitt, and Lieutenant Beaumont —and 44 other ranks.

* Private Byrne was awarded the D.C.M. for his services.

The total casualties in the Battalion, including those of the officers and men who during these days were temporarily attached, amounted to 8 officers and 10 other ranks killed, 7 officers and 162 men wounded, 17 officers and 713 other ranks missing, and 7 officers and 31 men gassed.

Carey's Force was now holding the line east of Hamel, and the 2nd Battalion Royal Dublin Fusiliers and 2nd Battalion Royal Munster Fusiliers were placed in support in a sunken road leading out of the town.   What was left of the 1st Battalion of the Regiment was sent up and incorporated in the 2nd Battalion, and the 16th Division was reduced to *one Infantry Brigade* consisting of two battalions of Infantry (one 48th Brigade battalion and one 49th Brigade battalion), one Pioneer, and one Engineer battalion.

The orders given to the Battalion were to reinforce the front line in case of emergency, or, in the event of the enemy gaining a footing in the front line, to counter-attack.

There was some heavy enemy shelling on the 29th causing several casualties, and next morning, after an hour's bombardment, the enemy attacked about 11.30 a.m. in a very resolute manner, upon which the Battalion moved forward clear of the barrage which had been put down on the support position, and counter-attacking in company with the Munsters, succeeded in recapturing that portion of the front upon which the German infantry had gained a temporary footing—but the Battalion sustained heavy casualties.

With the exception of hostile artillery fire nothing of importance was noticed during March 31st, and here the Battalion remained until the night of April 3rd, when it was relieved by a battalion from the 14th Division and marched to the neighbourhood of Blangy Trouville, whence it proceeded by omnibus to Saleux and here found what was left of the Division re-assembling.   On the afternoon of the 4th the 16th Division entrained at Saleux for the Hallencourt West Area, on arrival at which the Battalion was billeted at Ramburelles.

Writing of this period an historian* says :—" On this evening

* Conan Doyle, Vol. V, p. 150.

several of those heroic units which had fought themselves to the last point of human endurance from the beginning of the battle were taken from that stage where they had played so glorious and tragic a part. The remains of the 39th, 16th and 66th Divisions were all drawn back for reorganization. It was theirs to take part in what was a defeat and a retreat, but their losses are the measure of their endurance, and the ultimate verdict of history upon their performance lies in the single undeniable fact that the Germans could never get past them. Speaking of these troops an observer remarked : ' they had been fighting for nine days, but were very cheerful and full of vigour.' The losses of some units and the exertions of the individuals who composed them can seldom have been matched in warfare."

The five divisions composing the Corps had lost 25,000 men and 135 guns, 27 of these being heavy artillery.

The enemy continued his offensive for a very few days longer, but it may be said that on April 5th the last great battle of the Somme came to an end ; during its progress, and at the moment when the situation seemed for the Allies at its blackest, Marshal Ferdinand Foch had assumed supreme direction of the operations of the Allied Armies on the Western Front.

About April 10th the Brigade moved to Campagne-les-Boulonnais, and on the 14th to Cléty, and while here the two Regular Battalions of the Royal Dublin Fusiliers were formed into one composite battalion, two companies being contributed by each and the 2nd Battalion finding headquarters and staff ; Brevet Lieutenant-Colonel K. Weldon, D.S.O., assuming command and the amalgamated unit being described as the 1st/2nd Royal Dublin Fusiliers. On the morning of the 15th the Composite Battalion marched to Boeseghem, where it was busily employed in preparing a defensive position in front of Thiennes, the headquarters, staff and transport of the original 1st Battalion proceeded to Wavrans. Three days later—on the 18th—the amalgamation lately effected was ordered to be amended, and it was now given out that the 1st and 2nd Battalions of the Royal Dublin Fusiliers were to make up a new 1st Battalion of an establishment of 940 non-commissioned officers and men, commanded by Lieutenant-Colonel Moore, D.S.O., with the 1st Battalion staff; while the headquarters of the 2nd Battalion was to be reorganized as a Battalion Training Staff with 10 officers and 43 non-commissioned officers and men, all the 16th Division units reorganizing on similar lines.

On April 26th the 1st Battalion left the 16th Division, being reposted to the 29th Division.

The 2nd Battalion remained in and about this area, employed as a training battalion until the end of May, when, at a strength of 8 officers and 41 non-commissioned officers and men, it was taken out of the 48th Infantry Brigade of the 16th Irish Division, and moved by lorry to Requinghem, becoming one of the units of the 94th Brigade, 31st Division, preparatory to again serving as an ordinary battalion.

On June 2nd the Commanding Officer received the following letter from Brigadier-General F. W. Ramsay, C.M.G., D.S.O. Commanding 48th Brigade :—

" I can't tell you how much I feel the loss of your Battalion ; for two and a half years now I have had Dublin Fusiliers under my command, and I am very proud to have had the honour of having them in my brigade so long.   No troops could have served better and I am indeed grateful to them for all the good work they have done for me.

" I shall be most interested to hear how you get on.

" The best of good luck to you and the Battalion."

The 94th Brigade contained the 2nd Loyal North Lancashire Regiment, the 2nd Royal Munster Fusiliers and the 2nd Royal Dublin Fusiliers.

On June 6th the 7th Battalion Royal Dublin Fusiliers, which, since July, 1915, had served in Gallipoli, Serbia, Macedonia and Palestine, arrived in France and joined the 94th Brigade, when the greater numbers of its officers, non-commissioned officers and men were absorbed into the 2nd Battalion ; nearly all ranks were found

AMIENS AND ALBERT

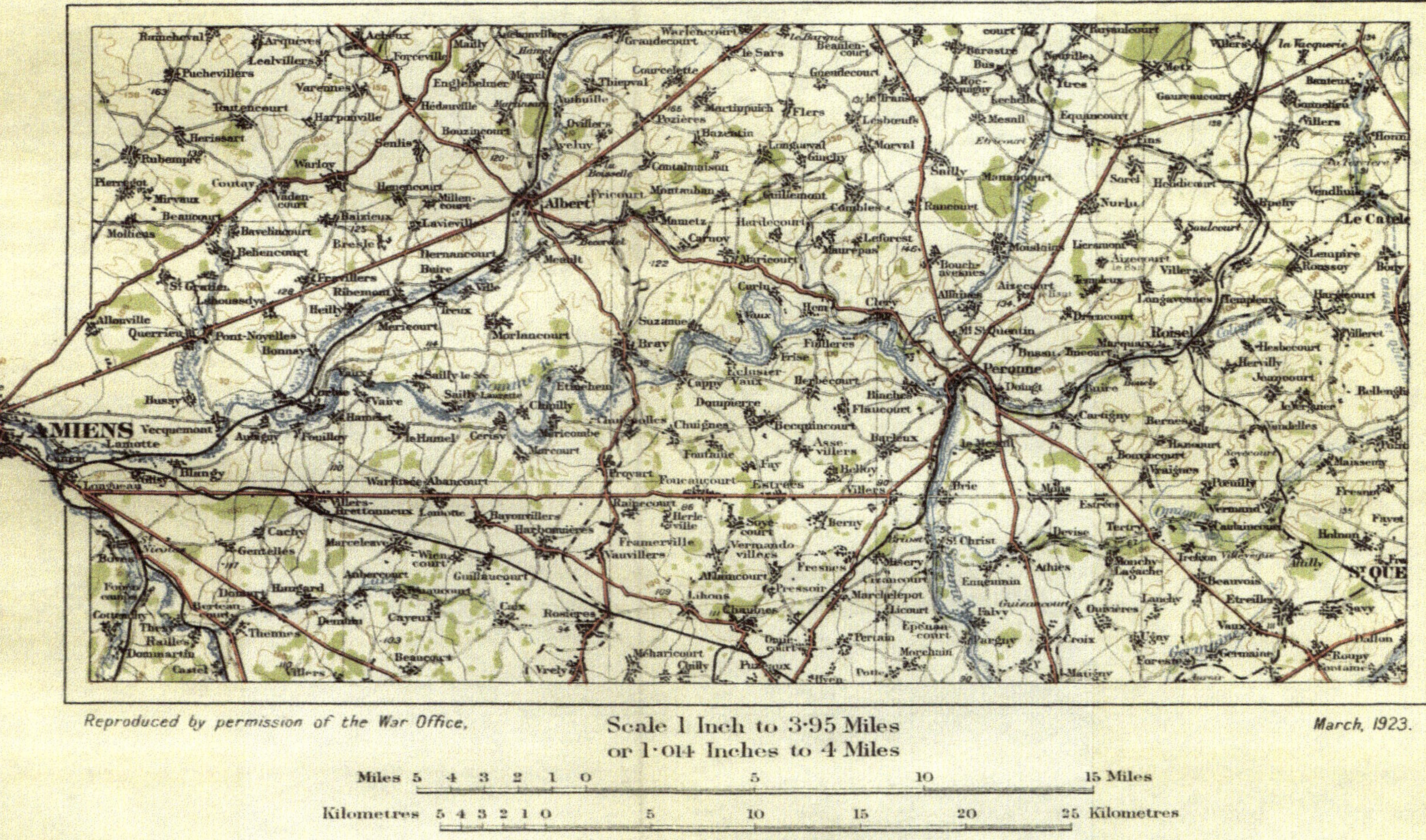

to be suffering more or less from malaria contracted in the Struma Valley.

On June 8th the Battalion marched to Val-de-Lumbres and on the 14th the officers were all sent by omnibus to Wallon Cappel and reconnoitred the army line of trenches near Hazebrouck with a view to possible occupation in the near future ; but if any such idea had been entertained it seems, by reason no doubt of the impaired health of a large proportion of the rank and file of the Battalion, to have been quickly abandoned, and the Brigade moved on the 15th by lorry and omnibus to Wizernes, where the train was taken and the Brigade finally occupied a camp at Rouxmesnil near Dieppe, where sea-bathing was enjoyed and training was seriously taken in hand. But on June 26th the 2nd Royal Dublin Fusiliers were again on the move and by the 28th were established in a new camp between the forest of Arques and the village of Martin Eglise.

The Battalion was now in the 149th Brigade, commanded by Brigadier-General P. M. Robinson, C.M.G., of the 50th Division, of which Major-General H. C. Jackson, D.S.O., was the commander, the other battalions of the Brigade being the 3rd Royal Fusiliers, the 1st King's Own Yorkshire Light Infantry and the 2nd Royal Munster Fusiliers.

On July 14th the Battalion—15 officers and 416 non-commissioned officers and men—paraded at 7 a.m. and marched to Dieppe, where the troops were drawn up in line on the *plage* in celebration of " France's Day." The massed bands of the 50th Division and Belgian buglers supplied the music. General Jackson took the salute and the regiments marched past ; on returning to camp the Battalion was met by General Sir Henry Rawlinson, the Army Commander, who complimented Lieutenant-Colonel Weldon on the smart appearance of all ranks.

During the month of July certain reinforcements arrived in camp and joined the Battalion ; on the 7th Lieutenants D. C. A. Shepard, C. W. Kidson, Second-Lieutenants S. A. Morris and S. Byrne arrived from the Base ; on the 8th Lieutenant G. R. Attwood was attached from the 7th Royal Irish Fusiliers ; on the 14th a draft of 30 men arrived from Calais ; Captain C. G. Carruthers, M.C., joined the Battalion on the 18th, followed next day by Captain A. Browne, 2nd Garrison Battalion Cheshire Regiment, and 60 men of the 7th Battalion Royal Dublin Fusiliers from Egypt ; while during the last ten days of July Lieutenant J. N. Barry and 35 other ranks joined the Service companies in camp.

At Martin Eglise the Battalion remained throughout August and the first half of September, going through all kinds of training and attending lectures on every possible military subject, until about September 14th the Brigade commenced to prepare for a move. On the 15th the Battalion entrained at Arques, arriving the same morning at Boquemaison, and marching from there to billets at Ivergny. It remained here until the 26th, when it proceeded by omnibus to Behancourt, where it was thus organized :—

> *Fighting Portion :* 24 officers and 612 non-commissioned officers and men.
>
> *Administrative Portion :* 2 officers and 91 non-commissioned officers and men.
>
> *Battle Surplus :* 9 officers and 181 non-commissioned officers and men.

The total strength was thus 35 officers and 884 non-commissioned officers and men. The fighting and administrative portions were under the command of Major L. C. Byrne, M.C., *vice* Lieutenant-Colonel Weldon who had proceeded on leave to the United Kingdom, and these now left Behancourt at 1.30 on the afternoon of the 28th for Nurlu, which was reached the same night and here the Battalion bivouacked. The Battle Surplus marched to Poulainville and occupied billets.

On the 30th the 2nd Battalion Royal Dublin Fusiliers was warned for immediate service, whereupon all battle stores were completed, fighting order was adopted, and officers and other ranks stood by ready to move up to the line at one hour's notice.

While the Battalion, and the Brigade and Division of which it formed part, had been out of the line, there had been almost continuous fighting on the allied front in the west. The Battles of

the Lys had endured from April 9th to 29th ; the Battles of the
Aisne and Marne of this year commenced at the end of May and
went on until the beginning of August ; while the Battle of Amiens
which immediately followed them, and the Battles of the Somme
which accompanied this last action, did not terminate until September 3rd ; while it was for the final breaking of the so-called
Hindenburg Line, leading to the battles which definitely established
the success of the Allies, that the 50th Division was now recalled
to the front.

" After the success of the British attacks on September 18th
and of the American attack on the St. Mihiel salient on September
12th, it had been decided between Marshal Foch and Sir Douglas
Haig that four divergent and simultaneous offensives should be
launched by the Allies—one by the Americans west of the Meuse in
the direction of Mezières ; the second by the French west of the
Argonne, in close conjunction with the American attack and in the
same direction ; the third by the British on the St. Quentin—Cambrai
front in the general direction of Maubeuge ; the fourth by the Belgian
and Allied forces in Flanders in the direction of Ghent. The most
important and critical of these attacks was the one to be undertaken by the British Armies against the Hindenburg Line."*

In consequence of the above, the Fourth Army (General Sir
Henry Rawlinson) was reinforced by the XIIIth Corps (Lieutenant-
General Sir T. Morland), with the 25th, 50th, and 66th Divisions,
and by the IInd American Corps of two divisions, and was given the
following orders :—" The Fourth Army, protected on its right flank
by the First French Army, will deliver the main attack against the
enemy's defences from Le Tronquoy to Le Catelet, both inclusive,
operating in the direction of the general line Bohain—Busigny.
The bombardment will commence on ' Z ' day (September 27th),
and the assault will be delivered on ' Z '+2 day (September 29th)."

On the night of October 1st the 149th Brigade relieved the
18th Division in the front, and next morning the Battalion moved
up to Tetard Wood, preparatory to an attack upon the Beaurevoir

* Montgomery, *The Story of the Fourth Army*, p. 137.

I

Line. On the 3rd the 50th Division was assembled with the 151st Brigade on the right, while the 149th on the left had two of its battalions holding a defensive flank along the St. Quentin Canal and one in support near the Knob ; the 150th Brigade was in divisional reserve. The 151st Brigade was the only one of the Division which this day appears to have been seriously engaged, and as a result of the day's fighting the Division was by 7 p.m. firmly established north of Gouy and Le Catelet.

On October 4th it had been intended that on the XIIIth Corps front the 25th Division on the right should capture Beaurevoir, while the 50th on the left was to seize the high ground north of Gouy and Le Catelet, between La Pannerie South and Richmond Quarry ; but the attack, which began at 6.10 a.m. in a thick fog, was not wholly successful, the 25th Division not being strong enough for the task assigned it ; and though the 50th gained possession of La Pannerie South, it was for some time held up by machine-gun fire from Hargival Farm and Richmond Copse. Late in the afternoon the 149th Brigade captured Hargival Farm, and a line was established along the northern slopes of Prospect Hill through La Pannerie South to Hargival Farm.

" D " Company of the Battalion attacked Maclincourt Farm at 4 p.m., but was driven off with four men missing. This day Battalion Headquarters moved up to Lone Tree Trench.

The attack on Beaurevoir having failed, it was decided to continue the attack on the 5th, and to endeavour also to capture the high ground between La Sablonnière and Guisancourt Farm. The attack by the 149th Brigade commenced at 5.45 a.m., " C " and " D " Companies of the Battalion advancing on the Hindenburg Line and taking Maclincourt Farm, in the capture of which Lieutenant Shepard was wounded. The remainder of the day passed quietly, and then at 3 on the morning of the 6th the attack was resumed, Hargival Farm and Quindampax Mill being assailed ; but unfortunately our troops were caught by our own barrage in No Man's Land and were held up, though later the objectives were gained and some 200 prisoners taken. The Brigade then pushed on without opposition to La Terrière, where it was relieved by the 33rd Division, and, with the 151st Brigade, was withdrawn into rest. The Battalion fell back for the night to Lone Tree Trench, marching on the 7th to La Pannerie, where it took up a position in trenches in support.

Of the result of these operations so far as they had gone, Sir Douglas Haig wrote in his Victory Despatch :—" The enemy's defences in the last and strongest of his prepared positions had been shattered. The whole of the main Hindenburg Line passed into our possession, and a wide gap was driven through such rear trench systems as had existed behind them. The effect of the victory upon the subsequent course of the campaign was decisive. The threat to the enemy's communications was now direct and instant, for nothing but the natural obstacles of a wooded and well-watered countryside lay between our armies and Maubeuge."

" On the Fourth Army front," writes Major-General Montgomery, " our troops had now reached open country, where the enemy had no prepared lines of defence, and which bore few traces of the devastation of war. It consisted of open undulating country devoid of hedges and free from wire, and was well suited to the employment of cavalry and tanks. The probable points of resistance, until the Selle was reached, were the villages, the small scattered woods north of Brancourt-le-Grand and Bohain, and the line of the railway running north and south a short distance west of Bohain and Busigny."

For the resumption of the attack, first fixed for the 7th and then postponed to October 8th, the XIIIth Corps on the left was to seize a line which included Le Hamage Farm and Les Marliches Farm as its first objective, joining up with the Americans on the right and the Third Army on the left ; the XIIIth Corps had attached to it one company of Whippets of the 3rd Tank Battalion, and the 1st Mark " V " Tank Battalion.

During the late afternoon of the 7th the Battalion had moved forward a short distance, but in so doing came under heavy shell fire and had several casualties, Lieutenants Sutherland and Elvey being wounded, the former mortally, while Corporal Noakes, who was killed, was recommended for the Victoria Cross, in that, although wounded, he went steadily on with his work of cutting gaps in the enemy wire.

This night the 149th Brigade was in support behind Vauxhall Quarry, and on the 8th was in divisional reserve, the Battalion moving up to Guisancourt Farm for the night, and was then on the 9th brought back to billets in Gouy; finally on the 10th proceeding by bus and march route via Maretz to Maurois, where on the 11th the Battalion was presented with a Tricolor flag by the Curé of Maurois; during the rest of the campaign this flag, with " R.D.F." emblazoned on it, was always carried as the Battalion Head-quarters flag.

At 3 p.m. on October 12th the Brigade moved to Honnechy, sustaining some casualties *en route*, Captain Pedlow being killed and Lieutenants Boulter and Poulter wounded, the former dying of his wounds at Roisel.

The interval between the above date and the 17th was spent in completing preparations for the attack now to be made by the Xth and XIIIth British and IInd American Corps. General Mont-gomery describes as " ambitious " the objectives given to the XIIIth Corps; they " comprised the capture of the whole of the ground lying between the Selle and the Sambre and Oise Canal, bounded on the south by the boundary with the French, and on the north by the Richemont River and the Bazuel—Catillon road, along which a defensive flank facing north-east was to be established. Le Cateau itself was included in the objectives."

On the 14th Second-Lieutenant Horrell and three scouts of the Battalion reconnoitred the River Selle to the east and south-east of St. Benin for fording places; on the 15th battle positions were reconnoitred, compasses checked, and all matters discussed at conferences; and at 5 p.m. on the 16th the Battalion—20 officers and 458 other ranks—left Honnechy and marched up to relieve the 2nd Battalion Royal Munster Fusiliers in the front.

The attack of the XIIIth Corps was to be carried out by the

50th Division on the right and the 66th on the left, with the 25th in reserve, the Honnechy—Le Cateau road being the dividing line between the two attacking divisions. The 151st Brigade of the 50th Division was to cross the Selle just north of St. Souplet ; the 149th was then to pass one battalion over the crossings made by the leading brigade and another by the demolished bridge at St. Benin, then moving on to the capture of the second objective, while the third battalion, following the right battalion of its brigade, was to turn north after crossing the railway and move on the railway triangle south-east of Le Cateau. The 150th Brigade was thereafter to pass through the 149th and capture Bazuel. The initial front of attack of the 50th Division was only 600 yards in width.

At 6.30 a.m. on the 17th the 2nd Battalion Royal Dublin Fusiliers crossed the Selle, " B " Company by a bridge which had been constructed, and " A," " C " and " D " Companies by fords. A very thick fog prevailed, but the companies were almost at once met by heavy machine-gun fire from which " A " Company especially suffered severely, the company commander, Lieutenant C. W. Kidson, and several men being killed and a number of officers and other ranks wounded ; the machine-gun fire seemed to come from a sunken road and from the railway embankment both parallel to the river and 100 and 800 yards distant from it respectively.

The 149th Brigade now attacked with three battalions in line, the 3rd Royal Fusiliers on the right, 2nd Royal Dublin Fusiliers in the centre, and the Scottish Horse on the left, but about 9.30 a.m., owing to the 151st Brigade having been checked short of its objective by a very obstinate resistance, the battalions of the 149th Brigade, instead of being able to proceed to the capture of the second objective, became embroiled in the fight, and while the 3rd Royal Fusiliers moved up in rear of a battalion of the 151st Brigade and formed a defensive flank facing south, the 2nd Dublins advanced to support another battalion of the 151st Brigade. The Battalion thus became intermingled with the 1st King's Own Yorkshire Light Infantry and these were held in check by the enemy in the orchards on the Arbre Guernon—Le Cateau road. At this time two strong counter-attacks were made by the enemy, and some units of the 50th British and 27th American Divisions were forced back.

As a result of this the Division naturally became somewhat disorganized and General Jackson now divided his command into three sections, the Battalion being in the centre section with the 1st Yorkshire Light Infantry and the 2nd Munsters, and holding the line from the orchards to the brick-works. The whole line was then firmly established and by 8 p.m. the brick-works were captured, but fighting continued in this part of the field throughout the night.

Early on the morning of the 18th the 50th Division—still in the three groups into which on the previous evening it had been divided—pressed forward to, first, the capture of the ridge 2,000 yards east of the Arbre Guernon—Le Cateau road ; second, the village of Bazuel ; and, third, the line of the Bazuel—Baillon Farm road. " At 5.30 a.m. the attack of the 50th Division was launched and was most successful from the outset. The first objective . . . was captured by the three groups of the division without much opposition . . . . A party of the 2nd Royal Dublin Fusiliers, over-running their objective, even penetrated into Bazuel and captured a few prisoners. Here, a daring individual exploit by Sergeant Curtis of this battalion put out of action the teams of two hostile machine-guns and resulted in the capture of four other machine-guns with their crews."*

The following is the official account of Sergeant Curtis's act in the *Gazette* awarding him the Victoria Cross :—

" No. 14017 Sergeant Horace Augustus Curtis, 2nd Battalion The Royal Dublin Fusiliers (Newlyn, Cornwall).

" For conspicuous gallantry and devotion to duty east of Le Cateau on the morning of October 18th, 1918. During an attack on an enemy position his platoon came unexpectedly under the intense hostile fire of many machine-guns. Knowing that the attack would be a failure unless the enemy guns were silenced, Sergeant Curtis without hesitation rushed forward through our own barrage and the enemy machine-gun fire. He reached the enemy

* Montgomery, p. 228.

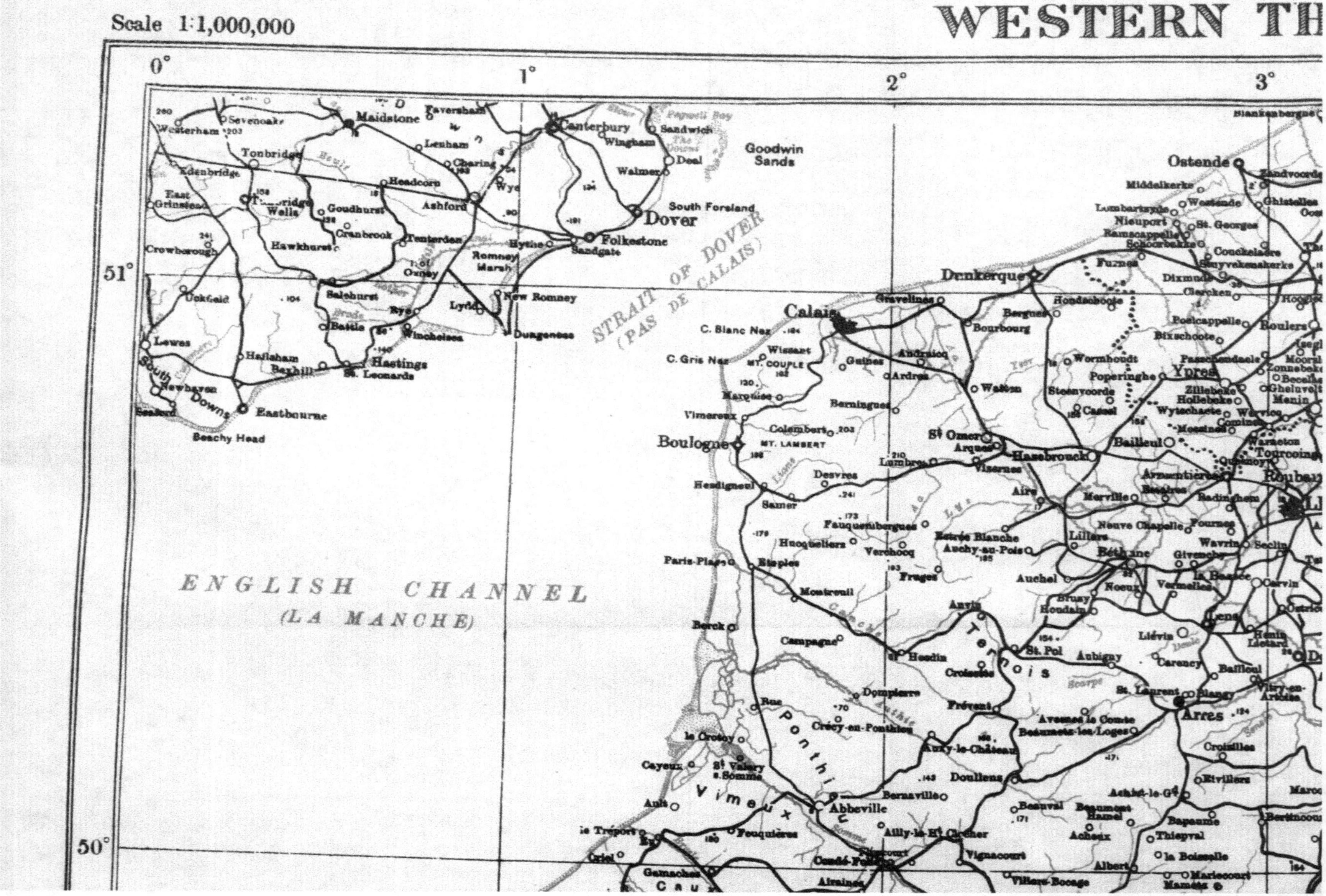

Scale 1:1,000,000

WESTERN TH

ENGLISH CHANNEL
(LA MANCHE)

STRAIT OF DOVER
(PAS DE CALAIS)

Goodwin Sands

Sevenoaks
Westerham
Tonbridge
Edenbridge
East Grinstead
Crowborough
Uckfield
Lewes
Newhaven
Seaford
Eastbourne
Beachy Head
Hailsham
Bexhill
Hastings
St. Leonards
Battle
Salehurst
Rye
Winchelsea
Hawkhurst
Cranbrook
Goudhurst
Tunbridge Wells
Headcorn
Maidstone
Faversham
Lenham
Charing
Ashford
Tenterden
Wye
Hythe
Romney Marsh
I. of Oxney
Lydd
New Romney
Dungeness
Sandgate
Folkestone
Dover
South Foreland
Canterbury
Wingham
Sandwich
Deal
Walmer

Ostende
Middelkerke
Westende
Zandvoorde
Ghistelles
Lumbartzyde
Nieuport
St. Georges
Ramscappelle
Couckelaere
Schoorbakke
Furnes
Dixmude
Clercken
Dunkerque
Gravelines
Bergues
Hondschoote
Bourbourg
Poelcappelle
Roulers
Bixschoote
Calais
C. Blanc Nez
Wissant
C. Gris Nez
MT. COUPLE
Guines
Andrsicq
Ardres
Watten
Wormhoudt
Steenvoorde
Cassel
Poperinghe
Ypres
Zillebeke
Hollebeke
Gheluvelt
Menin
Wytschaete
Messines
Warneton
Comines
Marquise
Vimereux
Berninghes
Colembert
Boulogne
MT. LAMBERT
St. Omer
Arques
Hazebrouck
Bailleul
Armentières
Roubaix
Desvres
Lumbres
Wizernes
Aire
Merville
Neuve Chapelle
Fournes
Hesdigneul
Samer
Fauquembergues
Estrée Blanche
Auchy-au-Bois
Lillers
Wavrin
Seclin
Hueqnelierz
Verchocq
Frages
Béthune
Givenchy
la Bassée
Paris-Plage
Etaples
Auchel
Bruay
Houdain
Noeux
Vermelles
Carvin
Montreuil
Anvin
Ternois
St. Pol
Liévin
Lens
Hénin Liétard
Berck
Campagne
Hesdin
Croisilles
Aubigny
Carency
Bailleul
Rue
Dompierre
Frévent
Avesnes le Comte
Beaumetz-les-Loges
St. Laurent
Blangy
Arras
Croisilles
le Crotoy
Crécy-en-Ponthieu
Auxy-le-Château
Etvillers
Cayeux
St. Valery s. Somme
Ponthieu
Doullens
Achiet-le-Gd
Ault
Vimeux
Abbeville
Bernaville
Beauval
Beaumont Hamel
Bapaume
le Tréport
Eu
Feuquières
Ailly-le-Ht Clocher
Acheux
Thiepval
Criel
Condé-Folie
Vignacourt
Albert
la Boisselle
Gamaches
Caux
Airaines
Villers-Bocage
Mametz
Maricourt

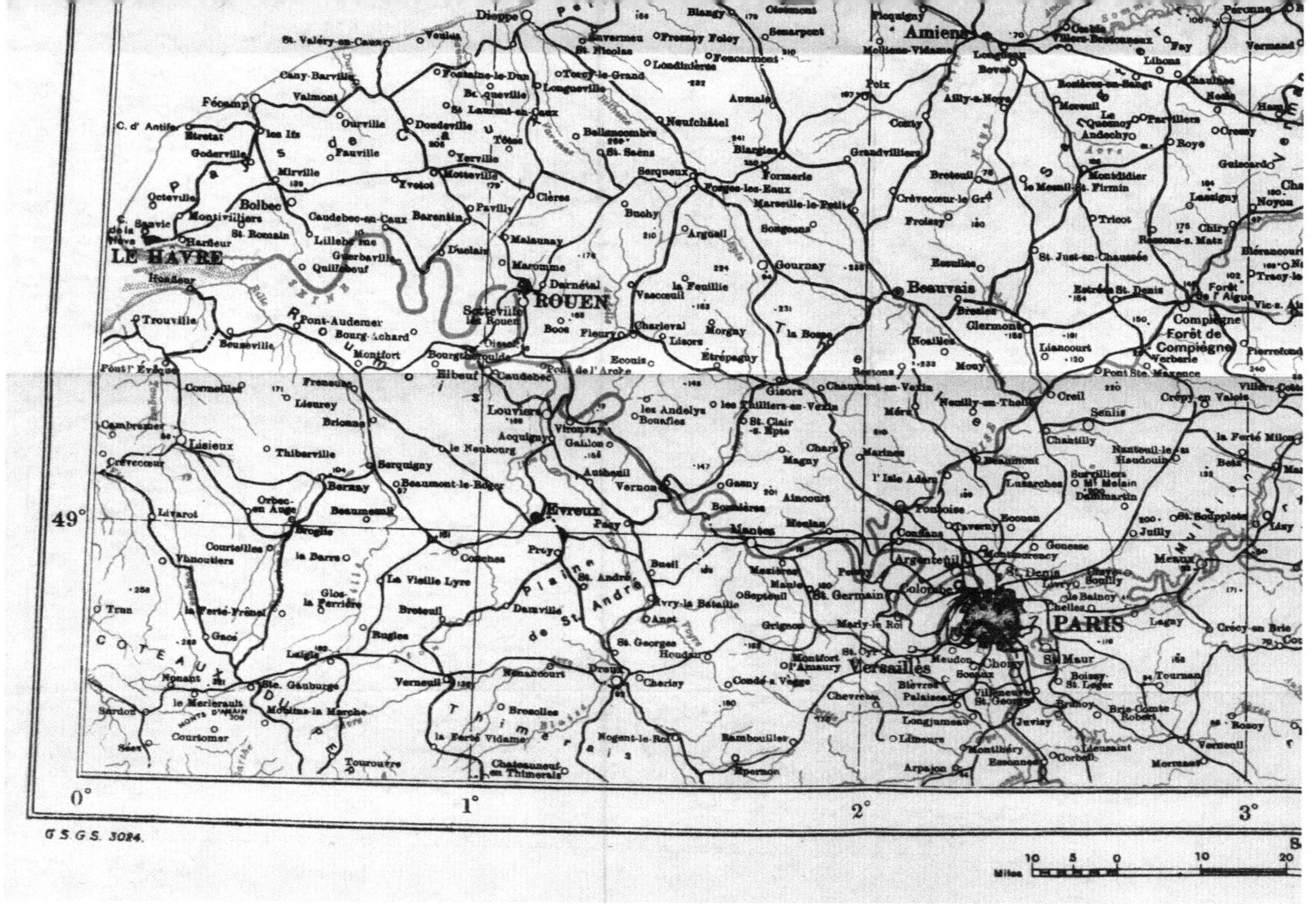
LE HAVRE
ROUEN
PARIS
Amiens
Beauvais
Versailles
St. Germain
Evreux
Lisieux
Dieppe
Fécamp
Bolbec
Neufchâtel
Gisors
Gournay
Mantes
Vernon
les Andelys
Pont-Audemer
Trouville
Bernay
Dreux
Argenteuil
St. Denis
Compiègne
Forêt de Compiègne
Senlis
Chantilly
Creil
Pontoise
Crépy-en-Valois
Brie-Comte-Robert
St. Valéry-en-Caux
Cany-Barville
Valmont
Fontaine-le-Dun
Bacqueville
Longueville
Torcy-le-Grand
Blangy
Fresnoy Folcy
Senarpont
Aumale
Poix
Ailly-s-Noye
Montdidier
Breteuil
Grandvilliers
Formerie
Forges-les-Eaux
Serquoux
Buchy
Argueil
la Feuillie
Charleval
Morgny
Étrépagny
Écouis
Pont de l'Arche
Elbeuf
Caudebec
Louviers
Vironvay
Gaillon
Aoquigny
le Neubourg
Acry
Brionne
Thiberville
Serquigny
Beaumont-le-Roger
Orbec-en-Ange
Livarot
Broglie
Conches
Authenil
Damville
Breteuil
La Vieille Lyre
Rugles
Laigle
Verneuil
Nonancourt
Anet
St. André
Ivry-la-Bataille
St. Georges
Houdan
Montfort l'Amaury
Rambouillet
Épernon
Chevreuse
Palaiseau
Longjumeau
Limours
Arpajon
Corbeil
Juvisy
Choisy
Sceaux
Meudon
Marly-le-Roi
Chelles
Lagny
Meaux
Gouesse
Taverny
Luzarches
Survilliers
Dammartin
Jully
Esbly
St. Just-en-Chaussée
Clermont
Liancourt
Pont Ste. Maxence
Verberie
Pierrefonds
Chaumont-en-Vexin
les Thilliers-en-Vexin
St. Clair-s. Epte
Magny
Chars
Marines
Beaumont
l'Isle Adam
Écouan
Gonesse
Colombes
Bréval
Bueil
Pacy
Gasny
Bonnières
Vernonnet
Bourgtheroulde
Montfort
Oissel
Sotteville-les-Rouen
Darnétal
Maromme
Malaunay
Pavilly
Barentin
Caudebec-en-Caux
Duclair
Guerbaville
Quillebeuf
Lillebonne
St. Romain
Montivilliers
Goderville
Étretat
C. d'Antifer
Harfleur
Honfleur
Yvetot
Motteville
Yerville
Doudeville
Tôtes
Bellencombre
St. Saëns
Blangies
Neufchâtel
Gaillefontaine
Marseille-le-Petit
Songeons
Crèvecœur-le-Grd
Froissy
Breteuil
le Mesnil-St. Firmin
Tricot
Noailles
Mouy
Mesnières
Estrées-St. Denis
Forêt de l'Aigue
G S G S 3024.
Miles
10 5 0 10 20
49°
0°   1°   2°   3°

THEATRE OF WAR

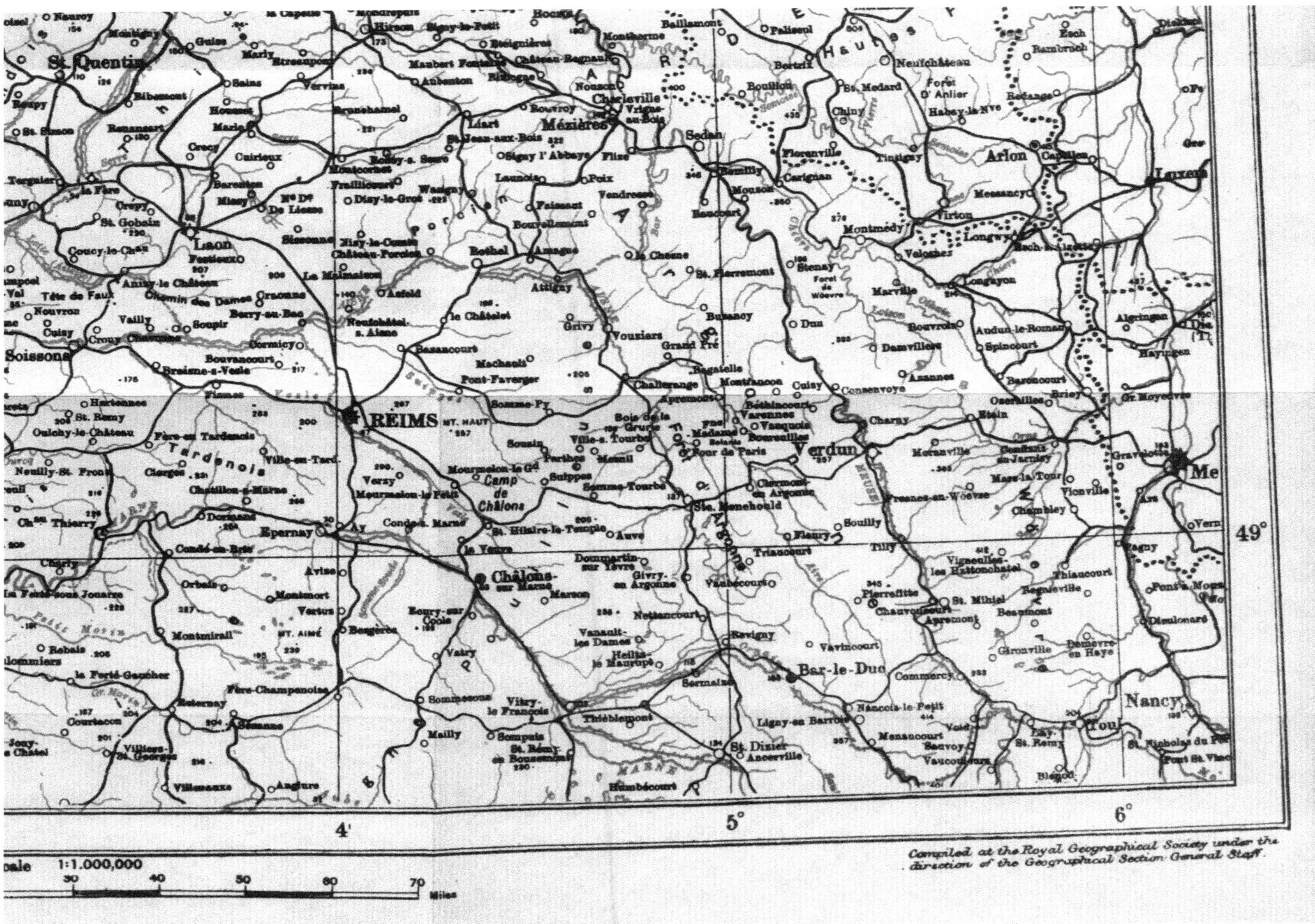
St. Quentin
Laon
Soissons
REIMS
MT. HAUT
Tardenois
Épernay
Châlons-sur-Marne
Montmirail
Montmort
Vertus
Fère-Champenoise
Vitry-le-François
St. Dizier
Ancerville
Bar-le-Duc
Verdun
MEUSE
Ste. Menehould
Argonne
Forêt de Paris
Vouziers
Grand Pré
Montfaucon
Sedan
Mézières
Charleville
Stenay
Dun
Montmédy
Longwy
Briey
Arlon
Luxem
Neufchâteau
Metz
Nancy
Toul
Commercy
St. Mihiel
Camp de Châlons
Mourmeron-le-Gd
Suippes
Vervins
Hirson
Guise
Marle
La Fère
St. Gobain
Coucy-le-Chât
Chemin des Dames
Craonne
Berry-au-Bac
Château-Thierry
Dormans
Avize
Chambley
Étain
Conflans
Thiaucourt
Pont-à-Mousson
Mars-la-Tour
Vionville
Fresnes-en-Wöevre
Beaumont
Vaucouleurs
Revigny
Sermaize
Triaucourt
Clermont-en-Argonne
Varennes
Vauquois
Compiled at the Royal Geographical Society under the
direction of the Geographical Section General Staff.
Scale   1:1,000,000
30   40   50   60   70   Miles
49°
4°   5°   6°

position and killed and wounded the teams of two guns. Through his extraordinary bravery and prompt action the teams of four other guns surrendered to him. A train-load of reinforcements was in the immediate vicinity, from which many of the enemy were detraining. He shot at the driver and succeeded in capturing over a hundred prisoners by the time his comrades reached him. His outstanding gallantry and disregard of personal safety inspired all near him to greater keenness and effort, which resulted in the attack on the whole battalion front being a complete success."*

The Battle of the Selle was now at an end so far as the XIIIth Corps was concerned, and had resulted in the capture by this corps of a carefully-prepared and strongly garrisoned position on a front of 7,000 yards, the greater part of which was protected by a difficult obstacle, while between the morning of the 17th and the evening of October 19th the Fourth Army had captured 5,139 prisoners, including 143 officers, and 60 guns.

The 50th Division now went into corps reserve, and on the 19th the Battalion was withdrawn and proceeded by way of Honnechy to Maretz, where it was accommodated in billets and set to work to reorganize and clean up.

Between October 16th and 18th the Battalion losses had amounted to a very high percentage of its battle strength ; as has already been stated the " Old Toughs " had gone into action with 20 officers and 458 non-commissioned officers and men, and of these 2 officers and 35 other ranks had been killed, 7 officers and 132 non-commissioned officers and men had been wounded, 24 men were gassed, there were five cases of shell shock and five men were missing —a total of 210 casualties in all ranks. The names of the officers are :—*Killed*—Lieutenant C. W. Kidson and Second-Lieutenant F. A. Walkey ; *wounded*—Captain A. B. Bagley, M.C., Lieutenants Humphrey, Byrne, Crawford, Staples, Lambkin and Glancy.

The captures made by the Battalion on October 17th amounted to 7 officers and 347 men, 9 trench mortars, 37 machine guns and 1 battery complete with teams.

* *London Gazette*, January 6th, 1919.

On October 25th Lieutenant-Colonel Weldon left the Battalion to assume temporary charge of the 54th Infantry Brigade, and, under orders of the Brigadier, the command of the Battalion was taken over for the time being by Major W. A. Trasenster, M.C., 3rd Battalion Royal Fusiliers.

By the beginning of November the Battalion was in bivouac at Pommereuil where on the 3rd a slight German bombardment with high bursting shrapnel caused a few casualties, Second-Lieutenant T. E. Flewett and one man being wounded, while another man was killed ; but next day a move was made at 3 a.m. to take part again in one of the major operations which was to bring the war to an end, the Brigade objective in this case being the southern portion of the Forest of Mormal.

The general attack now projected for the British Armies was to be delivered on a front of some 30 miles, from the Sambre and Oise Canal to Valenciennes, the general line of advance of the Fourth Army, on a frontage of about 15 miles, being due east. The difficulties of the country to be traversed were many and great ; there was the canal itself to be crossed at the outset, there was the canalised Sambre, the low ground had been inundated and was mainly swamp, while Mormal Forest covered an area of 40 square miles, and although much of the timber it contained had been cut down by the Germans the undergrowth in many parts was very thick and greatly hampered movement.

" The task of the XIIIth Corps entailed an attack through the southern portion of Mormal Forest, the forcing of the canal crossings at Landrecies, and a total advance of approximately 10 miles. . . . In view of the depleted strengths of his divisions and of the depth to which the advance was to be carried, Sir Thomas Morland decided to employ three divisions for the initial attack, each being on a comparatively narrow front, with one division in support. . . The 50th Division, in the centre, operating on a front of 2,500 yards, was responsible for clearing the portion of the Mormal Forest between the north boundary of the 25th Division " (on the right) " and a line drawn approximately due west from the bend of the canal 2,000 yards west of Sassegnies. The division was then to cross the canal and advance, in conjunction with the 66th Division, to the line of exploitation or second objective laid down by the Army Commander."*   This ran east of Cartignies, Dompierre and St. Remy Chaussée.

The 50th Division advanced at 6.15 a.m. on November 4th, the 149th Brigade on the right, the 150th on the left, and the 151st in support ; the 2nd Royal Dublin Fusiliers were in brigade reserve and moved forward from Fontaine-au-Bois 800 yards in rear of the attacking battalions, coming almost at once under a heavy barrage and sustaining many casualties. The machine-gun fire was also severe, the hedgerows being planted thick with these weapons, and the mist was dense and progress was consequently slow. When the attacking battalions, which had been temporarily checked along the Landrecies—Englefontaine road, had reached their objective in the Forest, the 2nd Battalion Royal Dublin Fusiliers was ordered to go forward and seize the spur which overlooks Landrecies and enfilades the Sambre Canal. This spur was attacked by " D " Company under Captain J. N. Barry and a few men of " C " Company led by Second-Lieutenant Morris, and was captured soon after mid-day, while an enemy field battery, attempting to come into action near the canal bank was also taken by the men of these companies, the teams and detachments being shot down and the battery commander made prisoner.

The guns taken were 4.2 howitzers, three 77 mm. guns and three machine guns.

All organized resistance now seemed to be broken and the only opposition encountered was from isolated machine guns firing at long range ; by dusk the 149th Brigade held the line of the canal from the bend north of Le Preseau to near Cense Toury.

The casualties during this day sustained by the Battalion were as follows :—*Killed*—Second-Lieutenant H. J. McBrien and 10 other ranks ; *wounded*—Captain A. M. Ewen, Lieutenants L. I. N. Lloyd Blood and G. H. McElnay, Second-Lieutenants T. B. Carrigg

* Montgomery, pp. 244, 245.

and F. V. Barry and 100 non-commissioned officers and men ; *missing*—3 men.

The Battalion now went for a very brief spell into billets and Lieutenant-Colonel Weldon rejoined on the 6th near Noyelles, where the Royal Dublin Fusiliers were at the time in support of the Royal Fusiliers and Scottish Horse who were then advancing on Monceau St. Vaast, which was captured the same evening.

On the 7th the Dublins marched to St. Remy Chaussée in close support to the same units of their Brigade, but no action ensued that day ; on the next, however, the 149th Brigade moved at mid-day by march route to Dourlers with orders to pass through the Division outpost line on the main Maubeuge—Avesnes road, and on arrival here the Battalion deployed through the line of the 4th King's Royal Rifle Corps, and proceeded to attack the enemy who was in occupation of Floursies. Major J. Luke, commanding the firing line, was soon wounded and the command of it then devolved upon Captain Kiernan.   There was considerable opposition from the German machine-gun nests, but this was satisfactorily dealt with and the Battalion was shortly in occupation of the village, but no captures in men or material were made.   At 4 p.m. an outpost line was thrown out beyond Floursies and defensive posts were placed on the flanks pending the arrival of other units whose advance had been delayed.

The casualties this day were Lieutenant Perrier and 4 men killed, Major Luke, Lieutenants Greaves and Lloyd Blood and 14 other ranks wounded.

This was the last action in which the 2nd Battalion Royal Dublin Fusiliers was engaged during the war, though early on the 9th the advance was resumed, the Battalion pushing through the wood without opposition while patrols advanced to Rue Haute ; but the enemy was in full retreat and had no longer the heart or the power to put up a strong resistance ; on the morning of November 10th the frontier of France was reached.

" The troops had been warned about 7 a.m." (on the 11th) " that hostilities were to cease at 11 a.m. The firing, however, which had been heavy all the morning continued until three minutes to 11 a.m. when it ceased for a short period and then broke out in a final crash at 11 a.m. The final act of a German machine-gunner, always our most formidable opponent throughout the war, is worthy of record. At two minutes to 11 a machine-gun, about 200 yards from our leading troops, fired off a complete belt without a pause. A single machine-gunner was then seen to stand up beside his weapon, take off his helmet, bow, and turning about walk slowly to the rear. Then all was silence. Combatants from both sides emerged from cover and walked about in full view."*

The World War was over.

In announcing on this day to Parliament the terms of the Armistice concluded with the representatives of the enemy nations, the Prime Minister spoke as follows :—

" Thus at 11 o'clock this morning came to an end the cruellest and most terrible war that has ever scourged mankind. I hope we may say that thus, this fateful morning, came to an end all wars. This is no time for words. Our hearts are too full of gratitude to which no tongue can give adequate expression. I will therefore move that the House do immediately adjourn until this time to-morrow, and that we, the House of Commons, proceed to St. Margaret's to give humble and reverend thanks for the great deliverance of the world from its great Peril."

And then, too, His Majesty the King, who had so often in the past four years, by the spoken word or the gracious message, expressed the feeling of his people towards his troops, his admiration of their valour and endurance, caused to be published the following last war-message to his Army :—

" I desire to express at once to all ranks of the Army of the British Empire, Home, Dominion, Colonial and Indian troops, my heartfelt pride and gratitude at the brilliant success which has crowned more than four years of effort and endurance.

" Germany, our most formidable enemy, who planned the war to gain the supremacy of the world, full of pride in her armed strength

* Montgomery, pp. 260, 261.

and of contempt for the small British Army of that day, has now been forced to acknowledge defeat.

" I rejoice that in this achievement the British forces, now grown from small beginnings to the finest army in our history, has borne so gallant and distinguished a part.

" Soldiers of the British Empire ! In France and Belgium the prowess of your arms, as great in retreat as in victory, has won the admiration alike of friend and foe, and has now by a happy historic fate enabled you to conclude the campaign by capturing Mons, where your predecessors of 1914 shed the first British blood. Between that date and this you have traversed a long and weary road ; defeat has more than once stared you in the face ; your ranks have been thinned again and again by wounds, sickness and death ; but your faith has never faltered, your courage has never failed, your hearts have never known defeat.

" With your allied comrades you have won the day.

" Others of you have fought in more distant fields ; in the mountains and plains of Italy ; in the rugged Balkan ranges ; under the burning sun of Palestine, Mesopotamia and Africa ; amid the snows of Russia and Siberia ; and by the shores of the Dardanelles.

" Men of the British race who have shared these successes have felt in their veins the call of the blood and joined eagerly with the Mother Country in the fight against tyranny and wrong. Equally those of the ancient historic peoples of India and Africa, who have learnt to trust the flag of England, hastened to discharge their debt of loyalty to the Crown.

" I desire to thank every officer, soldier and woman of our Army for services nobly rendered, for sacrifices cheerfully given ; and I pray that God, Who has been pleased to grant a victorious end to this great crusade for justice and right, will prosper and bless our efforts in the immediate future to secure for generations to come the hard-won blessings of freedom and peace."

[*Photo, Imperial War Museum.*

H.M. THE KING VISITS QUESNOY, DECEMBER 2nd, 1918.

CHAPTER VII

1918—1922.

DEMOBILIZATION AND RECONSTRUCTION.

CONSTANTINOPLE AND INDIA.

ON the day following the proclamation of the Armistice there was a Thanksgiving Service in the Chateau grounds at Dourlers, and then for many following days the Battalion furnished working parties for salvage work of all kinds, while training was commenced, ceremonial drill was once again practised, and classes of education were formed.

On December 1st His Majesty the King once more, as on so very many occasions in the course of the war, paid a visit to his Army in the Field, and, accompanied by the Prince of Wales and Prince Albert, inspected the troops of the 50th Division in a field off the Maubeuge—Avesnes road. At this date the Battalion was stationed at Baslieu, but on December 5th it moved by road to Monceau St. Vaast, where it was reorganized to meet the conditions created by the educational scheme which had been introduced and by the demobilization which was now impending.

Consequently, " A " and " D " Companies were formed of those non-commissioned officers and men who were employed in trade or had settled occupations prior to enlistment, and whose re-employment on return to civil life was already assured.

" B " Company took all serving soldiers and men attending school.

" C " Company was made up of those with or without trades who had no prospect of work on return to civil life, men learning trades in the Battalion under the new schemes which had been started, and those desirous of learning trades which the Battalion did not profess to teach, such as motor work, electric lighting and engineering of all kinds.

On December 18th a move was made, the Battalion leaving Monceau St. Vaast and marching to Le Quesnoy, where all ranks were gratified to learn the news of the award of the Victoria Cross to Sergeant Curtis.

During December a beginning was made of demobilization, 57 miners leaving the Battalion for home and discharge.

At the end of the year 1918 the Colours were sent out from home, and were handed over with befitting ceremonial to the Battalion on January 2nd, 1919, the Divisional Commander and Brigadier being present.

During this month demobilization set in and was very gradually carried out—thus on the 5th one man left, on the 7th 3 more, on the 15th an officer and 10 men took their departure, and so officers and men were sent away in small or large parties, until by the end of January the strength of the 2nd Battalion Royal Dublin Fusiliers had been reduced by 4 officers and 109 non-commissioned officers and men. During February the numbers demobilized dropped again, while men with two years or more Colour service to complete were allowed to go to the United Kingdom on twenty-eight days' leave of absence; but the demobilization orders initially announced were later changed, and as a not unnatural result hopes had been raised which it was not found possible to gratify, and some difficulty was experienced in making those men understand who had been unavoidably detained with the Colours, that there was as yet no peace and only a cessation of arms, that a state of war still existed, and that it would be necessary for some time at least to maintain an army in the field.

Long before the war came to an end the authorities had realized that very much more must be done than had ever before been attempted, to minimize the possible distress and certain confusion that must be occasioned by the sudden disbandment of hundreds of thousands of men possessing small means and having no immediate prospect of earning money. Still less could the ancient Israelitish

THE COLOURS, LE QUESNOY, 1919.

Left to right—Sergt. Downie, V.C., Lt. O'Sullivan, M.C., C.S.M. Walters,
Lt. Wolfe, M.C., Sergt. Curtis, V.C.

system of demobilization be pursued which followed when " the land rested from war " at the conclusion of the campaigns of Joshua and other leaders of the Jewish people, and when, as the Scriptures tell us, they simply " let the people depart, every man unto his inheritance."

So far back as January, 1915, a scheme was prepared for the consideration of the Cabinet of the United Kingdom, offering suggestions for meeting the difficulties likely to be experienced on the conclusion of the war by the release of the large numbers of men serving in the army on return to civil life. It was proposed to follow, as far as possible, the experience and precedents of earlier wars, while offering, in addition, a free insurance against unemployment, and it was arranged that certain existing organizations, and particularly the newly-created Labour Exchanges, should be made use of for finding suitable employment for demobilized men. These proposals were generally approved—in principle, but as there did not seem at the time any immediate prospect of their being carried into effect, they were provisionally pigeon-holed.

As the war progressed Committees were appointed to deal more in detail with demobilization, *e.g.*, a Reconstruction Committee, an Army Demobilization Committee, and others, and by these it was decided to give each man :—

1. A furlough, with pay and separation allowances for four weeks from date of demobilization.
2. A railway warrant to his home.
3. A twelve-months' policy of insurance against unemployment.
4. A money gratuity in addition to the ordinary Service gratuity.

Various alternative methods of dispersal were considered, and at first the principle was adopted and followed of granting release from army service in an order of priority determined by individual qualifications, with an eye at the same time to providing for a very early reconstruction of the Army after the declaration of peace. From the foregoing it will be realized that in considering and solving the problems of demobilization, the authorities had drawn up their scheme with a bias in favour of national interests, and that the army was to be dispersed in accordance with the requirements of the reconstruction of industry, and by individuals rather than by military units.

The scheme, as at first administered, met with much opposition both in the Army and in the Press of the country, while a system of " Special Releases," which formed no part of the original scheme, was justly open to the charge of " favouritism," and as a result Army Order No. 55 of 1919 finally abolished the principle of release on industrial grounds and substituted that of release on grounds of age and length of service.

Demobilization actually commenced in December, 1918, and proceeded with really remarkable regularity and dispatch ; before the end of February of the year following 1,848,000 men had been discharged, and of these 85 per cent had already returned to industrial occupations. At one time between 5,000 and 6,000 men were being daily returned to civil life : and on July 17th, 1919, the Secretary of State for War was able to announce that nearly 3,000,000 soldiers had been demobilized since the Armistice, leaving 1,200,000 still in the Army, including the 209,000 Volunteer Regulars.

In February, 1919, the Band of the 2nd Battalion Royal Dublin Fusiliers arrived from home, and on the 27th of this month a draft of 9 officers and 79 other ranks left the 2nd to join the 1st Battalion, then at Berg Gladbach in Germany ; the officers were Second-Lieutenants Morris, Gibbons, Buckley, Tully, Cooke, Woods, Conran, Horrell, M.C., and Coakley. This, and other defections, reduced so appreciably the strength of the Battalion, that on the last day of February it was reorganized in one company.

What was left of the 2nd Battalion remained on during March and April and May at Le Quesnoy, being more than once warned to leave France, but these orders being as often cancelled. However, at long last, on June 2nd the Cadre of the Battalion—five officers—Lieutenant-Colonel K. C. Weldon, D.S.O., Captain T. Brady, Lieutenant D. J. Davis, M.C., Second-Lieutenant E. D. A.

MAJOR & QUARTERMASTER J. BURKE,
D.S.O., M.C., D.C.M.

Artisani (Royal Fusiliers), and Major and Quartermaster J. Burke, D.S.O., M.C., D.C.M.—the only officer who had left England with the Battalion in 1914 and remained with it throughout, and 53 non-commissioned officers and men left Le Quesnoy by train for Le Havre and marched thence to camp at Harfleur. A stay was made in this camp until the 7th, when the Cadre marched down to the docks and embarked for Southampton in the *St. George*. Arrived here the officers and men disembarked and took the train for Colchester, which was reached at 9.50 p.m., and the Cadre was accommodated in the Foreign Service Detail camp at Read Hall Camp, where the Foreign Service Details of the Battalion had been formed since the beginning of April under the command of Major P. J. Shears.

Lieutenant-Colonel Weldon now went on leave, Lieutenant Davis was posted to the 3rd Battalion, while Second-Lieutenant Artisani rejoined his own corps.

The Cadre, with the Details under Major Shears, just reinforced by a draft of 151 non-commissioned officers and men from the 3rd Battalion, now began to assume something of the proportions of a battalion once more, and when on June 23rd, under command of Major Smithwick, it moved from Colchester to Aldershot, it was at a strength of 29 officers and 316 other ranks. At Aldershot it was accommodated in Badajoz Barracks, being posted to the Stanhope Lines Infantry Brigade under Brigadier-General Daly, C.B. Immediately after arrival here Major Conlan joined and took over command, but only held it for some three months, until October 21st, when he was relieved by Brevet Colonel C. Bonham-Carter, C.M.G., D.S.O., promoted from the Royal West Kent Regiment.

This winter H.I.M. the Shah of Persia paid a visit to England, and on the occasion of his being entertained at the Guildhall, the Battalion with band and Colours, strength 17 officers and 745 other ranks, assisted to line the streets, being employed first in the Strand and Fleet Street and later being stationed along the Victoria Embankment. A couple of days later the troops of the Aldershot Command paraded for inspection by H.I.M. when the following was published in Command Orders :—

K

" H.I.M. the Shah of Persia has desired General Sir Archibald Murray, C.-in-C., Aldershot Command, to convey to the troops which took part in the review on November 3rd, his sincere appreciation of their excellent turn-out and soldierly bearing. He was much struck with the march past of all the arms present."

On November 13th Major-General C. D. Cooper, C.B., Colonel of the Royal Dublin Fusiliers, visited the Battalion and inspected it, congratulating all ranks on the magnificent work done during the war of 1914–1918.

From the conclusion of the war a considerable British force had been maintained in Turkey, the strength of the British Army of Occupation in the Ottoman dominions on December 1st, 1919, being 13,000 British and 14,000 Indians, but of these a very large number were awaiting demobilization and it was necessary to provide replacements. At the end of November then the 2nd Battalion Royal Dublin Fusiliers received sudden orders for embarkation on December 3rd at Tilbury for Constantinople, but actually it was the 4th of that month before the Battalion left Aldershot and embarked in the hired transport *Rio Pardo*.

The Details remaining behind were placed under the charge of Captain W. H. Braddell and were attached to the 1st Battalion of the Regiment at Witley.

The following were the officers who sailed for Constantinople with the Battalion :—Brevet Colonel C. Bonham-Carter, C.M.G., D.S.O., in command ; Majors G. S. Higginson, R. L. H. Conlan, S. G. Smithwick, O.B.E., and D. French ; Captains A. S. Trigona, C. G. Carruthers, M.C., J. D. Glegg, M.C., C. W. Maffett and W. J. Shanks ; Lieutenants C. M. Craig-McFeely, D.S.O., M.C., W. H. Stitt, D.S.O., M.C. (adjutant), H. G. Aylmer, W. H. Hynes, H. M. B. Gun-Cuninghame, M.C., H. B. Harrison, M.C., L. A. Lawrence, D. C. A. Shepard, G. Petit, M.C., G. A. Hinkson, F. R. H. Macaulay, T. A. H. Chadwick, T. E. Flewett, D. H. Ross and M. A. Condron ; Second-Lieutenants M. H. FitzGerald, G. D. B. Russell and J. M. Allen (Education officer, attached), and Captain and Quartermaster F. W. Langley.

VICTORY MARCH DETACHMENT, JULY 1919.

[*Photo, Gale & Polden, Ltd.*

The embarking strength of the Battalion was 29 officers, 8 warrant officers, 40 sergeants, 40 corporals and 670 other ranks.

Gibraltar, where the ship remained 24 hours, was reached on the 11th and then, sailing eastward again, Stamboul was arrived at on the 20th, and the Battalion, disembarking at Haidar Pasha about 2.30 p.m., marched to the Turkish Medical School and Cavalry Barracks where it was to be quartered under the active service conditions here prevailing.

The General Officer Commanding the Allied Corps, Constantinople, was Lieutenant-General Sir H. F. M. Wilson, K.C.B., K.C.M.G., and the Battalion was posted to the 84th Infantry Brigade commanded by Major-General H. L. Croker, C.B., C.M.G., who, on December 30th, inspected the Battalion.

The Company Commanders at this time were :—" A " Company, Major Conlan ; " B," Major Smithwick ; " C," Major French ; and " D," Captain Trigona.

In accordance with the decision of the High Commissioners the formal military occupation of Constantinople and Scutari took place by the Allied forces on the morning of March 16th. This measure included taking over control of the Turkish War Office and Ministry of Marine, and the proclamation of martial law in Constantinople and Scutari. The Allied Navies closed the Bosphorus and the northern exit of the Sea of Marmora to all ferry and boat traffic ; and in the event of the conditions laid down by the Allies not being accepted by the Turkish authorities it had been decided to carry them out by force with the allied fleets and armies ; but the Turks did not prove contumacious and finally accepted the conditions as announced.

The following were the measures carried out by the Battalion :—

1. " C " Company formed an outpost line from the transport lines round the concentration camp and extending towards Kuchuk Chamlija and joining up with the 1st Battalion 2nd Rajputs of the Indian Army. The object of this was to cut off Scutari and prevent the passage of armed bodies or individuals to or from that place.

2. The occupation of Haidar Pasha telegraph and telephone offices and of Scutari telegraph office.
3. Active patrolling was arranged and maintained.
4. The arrest of several important Turkish leaders was successfully carried out.

On the night of March 16th—17th the outpost line was withdrawn and the patrolling ceased owing to the acceptance of the Allies' terms, but the guards were kept on for some further considerable time.

From April 2nd the Battalion came under the 84th Infantry Brigade for administration, but tactically was directly under the Headquarters 28th Division.

During May and part of June " B " and " D " Companies were stationed at Soghanli for training and musketry, the rest of the Battalion being at Haidar Pasha, but the guards and duties were so heavy that not much training could be performed. At the beginning of June the attitude of the Nationalist Party became noticeably more and more hostile. On the 15th information was received that the Nationalists had attacked the Anti-Nationalist Party which had been withdrawn and embarked at Ismid, and it was considered by no means improbable that the Nationalists might succeed in evading the British force at Ismid and so march on Scutari.

On the 16th then " A " Company, with two machine guns, occupied Biyuk and Kuchuk Chamlija in order to protect Scutari from the East, and three days later information came to hand that the Nationalist troops had commenced active hostilities against the British troops and were to be regarded as enemies. The evacuation of Soghanli was now decided upon, in view of the exigencies of the military situation, and " B " and " D " Companies accordingly rejoined headquarters at Haidar Pasha.

The first casualty sustained by the Battalion in the operations now commencing occurred on June 20th, when No. 21719 Private Brennan was killed in action with a band of Nationalist troops.

On the 28th the Battalion was ordered to vacate barracks and move to an advanced position near Alem Dagh, whereupon all heavy baggage was packed and stored, and on relief by the 2nd Battalion Essex Regiment, the Royal Dublin Fusiliers marched to Apostal Chift, six miles east of Haidar Pasha, and pitched camp, forming an outpost line for the protection of the Bosphorus—" D " Company occupying part of the line from Shaban Bair inclusive to Point No. 448. On July 3rd " A " and " C " Companies moved to Dodulu in relief of two companies of the 1st Battalion 54th Sikhs, the Battalion being now responsible for the portion of the line from Geuk Dagh inclusive to Point 448, the O.C. 2nd Battalion Royal Dublin Fusiliers being in charge of the left sector of the Brigade line.

A reconnaissance in force was carried out on July 5th towards Ermini Keui and the approaches thereto from the direction of Kara Bair and the open ground east of Ak Cheshme Sumari Karakol. The party, which was commanded by Colonel Bonham-Carter, was composed of one section 54th Battery, Royal Field Artillery, two companies of the Battalion, one company 95th Russell's Infantry, Indian Army, and one section of " Z " Machine Gun Company.

The enemy offered considerable resistance, and both sides suffered some casualties, the Turks losing 3 men killed, 3 wounded, and 18 captured, while there were 4 casualties in the 95th Infantry. Kara Bair was occupied, when the Nationalists retired in the direction of Laz Keui, and Colonel Bonham-Carter's troops searched Ermini Keui, the country about which was densely wooded and very hilly, the scrub and undergrowth being exceptionally thick and obliging the troops to keep to the tracks.

Colonel Bonham-Carter was now sent to take charge of the Beicos Defences, and Major Higginson assumed in his absence the command of the Battalion.

With a view to clearing the neighbouring country of the small bands of Nationalists, of whose presence reports came in from time to time, certain operations were now ordered to be carried out by the Battalion in conjunction with details, which included the 2nd Greek Battalion of the 4th Archipelago Regiment.

" A " Force, under command of Major Higginson, consisted of " C " and " D " Companies, 2nd Royal Dublin Fusiliers ; 1 company, 1st Battalion 2nd Rajputs ; 1 company, 54th Sikhs ; 1 section, 54th Field Battery; and 1 section, " Z " Machine Gun Company.

" B " Force contained 1 company and 1 platoon of the 25th Punjabis.

On the morning of July 23rd " A " Force took up a position extending from Sara Ghazi to Ermini Keui ; at this last named place the enemy offered a good deal of resistance, and during the day the village, occupied by " D " Company of the Battalion, was several times heavily fired on ; but the enemy lost 12 men killed, and in the evening a reinforcement was sent up of one company of the 54th Sikhs. The night passed quietly, and in the course of the morning " A " Force moved forward and occupied a line of hills to the south of Eumerli, with " B " Force on the right flank. Some opposition was met with in moving to this position, but it was quickly overcome, 3 of the enemy being killed. The position now occupied by the two companies of the Battalion overlooked the valley in which the village of Eumerli is situated.

It had been arranged that the Greek Battalion of the Archipelago Regiment was by now to have been in position on the north side of Eumerli, ready to advance at 11 a.m. against the village and drive the Nationalists, believed to be holding it, towards the position occupied by the Royal Dublin Fusiliers. But the Greeks did not come into line until three hours after the time arranged, and the enemy was thus able to escape in a north-easterly direction towards Shile. Some 200 rounds of gun ammunition were found in the village and destroyed. About 5 p.m. " A " and " B " Forces withdrew, the greater part of the former bivouacking for the night of July 24th—25th in the vicinity of Sultan Chift. The British forces suffered no casualties, while 15 of the enemy's dead were counted.

The outpost line was maintained until the month ended, on which date the Battalion numbered 28 officers and 827 other ranks ;

1914
LE CATEAU
THE MARNE

1915
YPRES
GALLIPOLI

1916
BEAUMONT
HAMEL
SOMME

1917
ARRAS
SALONICA

1918
GERMAN
OFFENSIVE
BRITISH ADVANCE
PALESTINE

1919
GERMANY

FOR HE TODAY WHO SHEDS HIS BLOOD WITH ME SHALL BE MY BROTHER

KING Henry V.
Act IV Scene III

SPECTAMUR AGENDO

THE ROYAL DUBLIN FUSILIERS

1914

1919

Presented to

by all ranks of
The Royal Dublin Fusiliers
to Commemorate your Service
with the Regiment Overseas
during the Great War, 1914-19,
when by your splendid determination, unfailing devotion to duty
and readiness at all times to
suffer the supreme sacrifice, you
nobly played your part in winning
through to Victory.

Colonel
The Royal Dublin Fusiliers

CARNATIC
MYSORE

PLASSEY
BUXAR

ARCOT • CONDORE
WANDIWASH • PONDICHERRY
GUZERAT • SHOLINGUR • NUNDY DROOG
AMBOYNA • TERNATE • BANDA • LUCKNOW
SERINGAPATAM • KIRKEE • MAHEIDPOOR • BENI BOO ALLI • AVA • ADEN
MOOLTAN • GOOJERAT • PUNJAUB • PEGU • R. of LADYSMITH • SOUTH AFRICA 1899-02

and then on August 3rd Headquarters, " A " and " B " Companies were at Dodulu, while " C " and " D " were in camp at Pasha Keui, whence active patrolling was for some little time longer kept up, but in no case was touch established with the enemy, and as the military situation gradually became easier, company and musketry training was resumed, until on August 30th the Battalion was relieved by the 2nd Battalion Essex Regiment ; Headquarters and the Right Wing then marched to Haidar Pasha, " C " and " D " Companies to Apostal Chiftlik, Colonel Bonham-Carter now returning to command.

On November 1st verbal orders were received that the 2nd Battalion Royal Dublin Fusiliers, less *personnel* due to proceed to the United Kingdom during the current trooping season, would embark for India about the 8th of the month in the hired transport *Field Marshal*. This date was not very closely adhered to, for it was not until the 17th that the Battalion actually embarked at Haidar Pasha. It was accompanied as far as Port Said by those details which were due for home, these being landed on the 22nd ; but a few other men, mostly those in hospital, remained behind temporarily with the " Army of the Black Sea." The embarking strength was 23 officers, 7 warrant officers, 30 sergeants, 8 lance-sergeants, 21 corporals, 47 lance-corporals, 408 privates and 6 boys. The following officers proceeded from Constantinople to India with the Battalion :—Brevet Colonel C. Bonham-Carter, C.M.G., D.S.O. ; Majors S. G. Smithwick, O.B.E., and D. French ; Captains A. S. Trigona, C. G. Carruthers, M.C., J. D. Glegg, M.C., and C. W. Maffett ; Lieutenants W. H. Stitt, D.S.O., M.C., C. M. Craig-McFeely, D.S.O., M.C., H. G. Aylmer, W. H. Hynes, H. B. Harrison, M.C., A. G. L. Sidwell, T. E. Flewett, T. A. H. Chadwick, D. S. Norman, M. A. Condron ; and Second-Lieutenant M. H. FitzGerald ; attached were Captains T. L. Henderson, R.A.M.C., and J. M. Allen, Education Officer, and the Rev. R. Barry Doyle, C.F.

Major G. S. Higginson, Lieutenant L. A. Lawrence and 154 non-commissioned officers and men were left behind at Port Said for conveyance to the United Kingdom.

At Suez the 16th Lancers were taken on board the *Field Marshal* which then proceeded on her voyage ; Karachi was reached on December 7th, when the Battalion at once disembarked and entrained for Multan. At this city—which the Battalion had first entered during the Second Sikh War and last visited at the close of the Great Mutiny—the train arrived in the early morning of December 9th, when the 2nd Royal Dublin Fusiliers relieved the 2nd Battalion of the Buffs in Edwardes Barracks, and found themselves in the 12th Indian Infantry Brigade, commanded by Brigadier-General A. J. Poole, C.M.G., and in the Lahore District under Major-General Sir S. T. B. Lawford, K.C.B.

The year 1920 which was now drawing to its close, had witnessed a very great change in the Battalion, wherein strenuous efforts had been made to form a well-trained, thoroughly disciplined and homogeneous unit out of the numbers of non-commissioned officers and men who had been collected together at Aldershot immediately prior to the departure of the 2nd Royal Dublin Fusiliers for Constantinople, just twelve months previously. The conditions prevailing in the Army of the Black Sea in which it there found itself included were in many respects those of active service ; the accommodation was very poor, anything like real comfort for the men was almost non-existent, and the many calls made for extra-regimental employ, coupled with the heavy duties, did not tend to make easier the work of re-forming the Battalion after four years of war. The country in which the Battalion passed the year was in a very unsettled state, demanding constant vigilance, while the many minor operations requiring to be undertaken against the Nationalist forces made it practically impossible to pursue the usual tenor of military training. The new system of educational training especially suffered and the results obtained were consequently very poor.

Much good and lasting work had, however, been effected in the Battalion branch of the National Savings Association started in November, 1919, and which had created something like a record among infantry battalions of the Army, the non-commissioned officers and men having invested over £10,000 in the course of the year.

During the year 1921 H.R.H. Field-Marshal the Duke of Connaught, Colonel-in-Chief of the Royal Dublin Fusiliers, visited India by command of His Majesty the King to open at Delhi the All-India Legislative Assembly and Council of State, and also the newly-constituted Chamber of Princes; and on the occasion of H.R.H. paying a visit to Rawal Pindi a guard of honour, provided by the Battalion, proceeded thither from Multan on February 14th; it was composed of Captain Carruthers, Lieutenants McFeely, Hynes, Harrison and Shepard, Band and Colour, 2 Colour-Sergeants and 100 rank and file. The following letter was received by the Commanding Officer from the Military Secretary to His Royal Highness :—

> " RAWAL PINDI,
> " *February 18th*, 1921.

" MY DEAR COLONEL,

" *Field-Marshal His Royal Highness the Duke of Connaught desires me to write to say what an unexpected pleasure it was to him to find a guard of honour from the Regiment provided for him at Rawal Pindi. He wishes me to say that he was very struck by their smart appearance and drill and he sends his best wishes to all ranks and trusts they may have a very happy sojourn in India.*

> " *Yours sincerely,*
> (*Signed*) " RIVERS WORGAN,
> " *M.S. to H.R.H.*"

During the hot weather of this year part of the Battalion was sent up to the stations of Kasauli and Solon in the Simla Hills.

On October 18th Major G. S. Higginson was appointed to command the Battalion in the place of Brevet Colonel Bonham-Carter, promoted to the command of the South Wessex Infantry Brigade in the United Kingdom; later in the month there was some trouble in the native city of Multan caused by the preachings of seditionists, and two companies of the Battalion were called out, but were not required to take any really serious action.

On November 4th medals and decorations for the Great War were presented on parade to those entitled to receive them ; and the following are the remarks made by H.E. the Commander-in-Chief on the inspection reports of the Battalion :—

" The progress that is being made is satisfactory."

Further, General Sir W. R. Birdwood, Bt., G.C.M.G., K.C.B., K.C.S.I., C.I.E., D.S.O., Commanding-in-Chief the Northern Command, after inspecting the Battalion on December 5th of this year, expressed to the Commanding Officer his " *admiration of the splendid turn-out and physique of the Battalion. Their general bearing and soldier like appearance were superior to that of any battalion seen up to date. He asked the Commanding Officer to congratulate all ranks on the good reports he had received regarding the behaviour of the Battalion since its arrival in India.*"

With effect from January 1st, 1922, the designation " Fusilier " replaced that of " Private " in the Royal Dublin Fusiliers.

On February 4th a telegram—which can hardly in view of all that was happening in the United Kingdom have been unexpected—was received from the Headquarters, Lahore District, announcing that the Battalion would move to England about the end of March ; and something must now be said to account for the sudden termination of a tour of Indian service which had little more than begun.

For many months past there had been very serious unrest in Ireland, amounting at times to actual rebellion against the forces of the British Government. Martial law had been proclaimed in one after another of the more disturbed districts, and this had gradually been extended until the rule of the military had almost everywhere superseded that of the Civil Power.

In June, 1921, His Majesty the King had paid a visit to Belfast and had there opened the Northern Parliament, while in the speech which he delivered on that occasion he made a very special appeal to all Irishmen to unite in securing peace for the distressed country ; and as the result of His Majesty's words the Prime Minister issued an invitation to the heads of the Northern Government and to the leaders of Sinn Fein to discuss matters with a view to a settlement

[*Photo, Bremner, Simla.*

"OLD TOUGHS" ON PARADE, MULTAN, MARCH 1922.

at a conference to be called in London.   Some sort of an agreement
was provisionally arrived at, and in July a truce was arranged
between the forces of the Crown and those of the Irish republican
party, pending the result of the negotiations then opening between
the British Government and the Sinn Fein leaders ; and a few days
later General Macready, commander-in-chief in Ireland, sanctioned
the removal of all restrictions on fairs, markets, etc., which had been
imposed in the martial law areas.

In the middle of July conversations were opened between
Messrs. Lloyd George and de Valera, and some three weeks later
the Government decided unconditionally to release all members of
the Dail Eireann, whether imprisoned or merely interned, in order
to permit of their attending the meeting of the Dail ; but after a
secret session of some few days' duration the Dail unanimously
rejected the Government's peace offer.   At the end of September,
however, the Premier again invited Sinn Fein to send representatives
to a fresh conference in London, and on the invitation being accepted,
negotiations opened on October 11th.   The situation was more
than once imperilled by certain declarations attributed to Mr. de
Valera, as when in a telegram to His Holiness the Pope, he dis-
claimed on behalf of his followers " allegiance to the British King,"
but the Conference went on and at last on December 6th an agreement was reached and its provisions were at once published.

The only one of these which immediately concerns this History
was that which affected the status and even the existence of such
Irish regiments of the British Army as were recruited in that portion
of Southern Ireland which, under the Treaty proposed, was for the
future to be known as " the Irish Free State " ; this agreed to the
establishment by the Government of the Irish Free State of a
Military Defence Force, the size of which was not to exceed such
proportion of the military establishments in Great Britain as that
which the population of Ireland bore to that of England, Scotland
and Wales.

But it was not only the future of Ireland which seemed likely
to affect the continued existence of the Irish regiments of the British

Army; in August, 1921, a committee had been appointed by the British Cabinet to advise on questions of finance, and to decide how to reduce the national expenditure for the ensuing year by no less a sum than one hundred and thirty millions. The committee had made proposals for very drastic economies in the Army, to be effected by disbanding many units, and it was generally assumed in the Service and other papers that, in view of the difficulty or indeed impossibility under the new regime of maintaining South Irish regiments in the British Army, it was practically inevitable that a large number of historic Irish regiments must disappear from the Army List.

For some little time after receipt of the warning telegram above mentioned, the Battalion pursued its ordinary manner of life; the drafts—though small in number considering the usual establishment of a British battalion in India—continued to be sent out from home, one 58 strong arriving in the beginning, and another of 56 men arriving at the end, of the month; while in this month also a detachment of 12 officers and 300 other ranks with band and Colours, under the command of Major French, proceeded to Lahore in connection with the visit to India of H.R.H. the Prince of Wales.

But with the issue of the Army Estimates the Secretary of State for War gave out that in order to effect the necessary economies demanded by the parlous condition of the finances of the country, it had been decided to reduce the Regular Army by five Line Cavalry regiments, or their equivalent, and twenty-four Line Battalions, and this decision was given effect to, so far as the Infantry of the Army was concerned, in Army Order No. 78, published on March 11th, and which is here given in full :—

"*Reduction of Establishment* :—His Majesty the King has approved with great regret the disbandment, as soon as the exigencies of the Service permit, of the following corps and battalions of Infantry of the Line :—

The Royal Irish Regiment, comprising :—
    1st Battalion.

[*Photo, Bremner, Simla.*

MULTAN, MARCH, 1922.

Back Row (left to right)—2/Lt. J. T. Thornton, I.A.U., Lt. D. C. A. Shepard, Lt. E. G. Utley, Lt. C. M. Craig-McFeely, D.S.O., M.C., Lt. G. W. C. Bolster, Lt. H. G. Aylmer, Lt. H. B. Harrison, M.C., Lt. T. A. H. Chadwick, Lt. M. H. FitzGerald, Lt. G. D. B. Russell.

Sitting—Capt. J. D. Glegg, M.C., Capt. G. W. B. Tarleton, M.C., Major D. French, Bt. Lt.-Col. J. McD. Haskard, C.M.G., D.S.O., Lt. Col. G. S. Higginson, Lt. and Adjt. W. H. Stitt, D.S.O., M.C., Capt. C. G. Carruthers, M.C., Capt. J. M. Mood, O.B.E., M.C.

On Ground—Lt. D. S. Norman, Lt. T. E. Flewett, Lt. M. A. Condron, Lt. A. G. L. Sidwell, Capt. C. W. Maffett.

2nd Battalion.
3rd Battalion (Militia).
4th Battalion (Militia).
Depot.
The Royal Irish Fusiliers (Princess Victoria's), comprising :—
1st Battalion.
2nd Battalion.
3rd Battalion (Militia).
4th Battalion (Militia).
Depot.
The Connaught Rangers, comprising :—
1st Battalion.
2nd Battalion.
3rd Battalion (Militia).
4th Battalion (Militia).
Depot.
The Prince of Wales's Leinster Regiment (Royal Canadians)
comprising :—
1st Battalion.
2nd Battalion.
3rd Battalion (Militia).
4th Battalion (Militia).
5th Battalion (Militia).
Depot.
The Royal Munster Fusiliers, comprising :—
1st Battalion.
2nd Battalion.
3rd Battalion (Militia).
4th Battalion (Militia).
5th Battalion (Militia).
Depot.
The Royal Dublin Fusiliers, comprising :—
1st Battalion.
2nd Battalion.
3rd Battalion (Militia).

4th Battalion (Militia).

5th Battalion (Militia).

Depot.

3rd Battalion, The Royal Fusiliers (City of London Regiment).

4th Battalion, The Royal Fusiliers (City of London Regiment).

3rd Battalion, The Worcestershire Regiment.

4th Battalion, The Worcestershire Regiment.

3rd Battalion, The Middlesex Regiment (Duke of Cambridge's Own).

4th Battalion, The Middlesex Regiment (Duke of Cambridge's Own).

3rd Battalion, The King's Royal Rifle Corps.

4th Battalion, The King's Royal Rifle Corps.

3rd Battalion, The Rifle Brigade (Prince Consort's Own).

4th Battalion, The Rifle Brigade (Prince Consort's Own)."

But already in the month previous—in February—it had been given out more or less officially that certain regiments and battalions were to be disbanded owing to the decision to make drastic reductions in the British Army, and that the Royal Dublin Fusiliers were to be one of the regiments thus to be sacrificed.

In a Special India Army Order it was now announced that four Irish battalions—" the 2nd Battalion Royal Irish Regiment, the 1st Battalion Connaught Rangers, the 1st Battalion Leinster Regiment, and the 2nd Battalion Royal Dublin Fusiliers—will proceed to the United Kingdom this trooping season—probably at the beginning of April—for disbandment," and giving the names of thirty-two battalions of infantry then serving in India to which warrant officers, non-commissioned officers and men of these Irish battalions might voluntarily transfer, failing which they would be discharged or transferred to the Army Reserve as their regiments were disbanded. As a result of this Army Order two warrant officers, twenty-four non-commissioned officers and one hundred and two Fusiliers transferred to units or departments in India.

In the case of officers these could, if they so desired, remain in India attached to other units pending further orders as to their disposal ; the following were the officers of the Battalion who elected to so remain : Lieutenants Craig-McFeely, Shepard, Flewett and Norman, attached to the 1st Battalion Border Regiment, and Lieutenants Dolan, FitzGerald and Russell to the 2nd Battalion York and Lancaster Regiment pending appointment to the Royal Air Force.

The following very sympathetic letter was received from General Lord Rawlinson, Commander-in-Chief in India, dated Delhi, March 8th :—

" DEAR HIGGINSON,—

" *It was with feelings of deep regret that I heard the official decision to abolish the Irish Regiments included the Royal Dublin Fusiliers.*

" *I realise to the full the sorrow with which this news must have been received by all ranks of your famous regiment, and I find it difficult adequately to express the sympathy I feel—a sympathy which is shared by the whole of the British Army.*

" *Your records of gallant service in every part of the world for nearly 300 years will go down to fame in history, and it is not for me to recall them ; but during my service it has often been my good fortune to be associated closely with your regiment, so I know it well.*

" *I recall with admiration the gallant services of your two Regular Battalions in the South African Campaign, and it has been my privilege to have under my command both Regular and Service battalions in the Great War.*

" *I well remember the gallant feat of arms performed by your battalion at the crossing of the Selle in October, 1918, which largely contributed to the success of the Fourth Army on that occasion.*

" *I very greatly deplore the necessity which entails the termination of your glorious record, though I feel it will never be forgotten.*

" *I would be glad if you would say a word of farewell on my behalf to all ranks and convey to them my heartfelt wishes for their prosperity.*

" *Yours very truly*
(*Signed*) " RAWLINSON."

Colonel Higginson replied as follows on behalf of the Battalion :

" MULTAN,

" March 11th, 1922.

" DEAR LORD RAWLINSON,—

" *I have to-day communicated to all ranks the contents of your letter of the 8th inst., and I desire to express on their behalf and my own, our thanks and grateful appreciation.*

" *The privilege of service under you, which dates back to Aldershot, 1909, is one of the Battalion's happiest recollections.*

" Yours very truly,

(Signed)   " G. S. HIGGINSON."

General Sir William Birdwood, Commanding-in-Chief, Northern Command, also telegraphed the message which here follows :—

" *Good-bye and all good wishes to all ranks of the gallant Dublin Fusiliers. I do indeed regret that such a magnificent battalion should be leaving my command, and more so knowing that I shall never have the honour of again serving alongside of you. I trust that all good fortune may be in store for you all.*"

To this kindly message Colonel Higginson made brief reply :—

" *Very many thanks on behalf of all ranks for your good wishes. Morituri te salutant.*"

It is very greatly to be regretted—though no doubt it would but have added to the pangs of dissolution—that the " Old Toughs " could not have sailed for home from the Port of Bombay, at which, more than two hundred and fifty-seven years previously, the regiment had first landed in India, but it was not to be.

On March 12th Lieutenant Bolster left Multan in advance with 14 married families and 30 other ranks *en route* to the United Kingdom, and then on April 5th the Battalion left its cantonments in two special trains for Karachi on the first stage of its final journey to disbandment at a strength of 18 officers, 10 warrant officers, and 652 other ranks, with 4 officers' and 7 warrant officers' families, embarking on April 6th in the *s.s. Assaye.*

MEMBERS OF SERGEANTS' MESS, 2ND BN. ROYAL DUBLIN FUSILIERS, MULTAN, MARCH 1922.

The following were the officers who accompanied the Battalion :
—Lieut.-Colonel Higginson, Brevet Lieut.-Colonel Haskard,
C.M.G., D.S.O., Major French, Captains Mood, O.B.E., M.C.,
Carruthers, M.C., Tarleton, M.C., Glegg, M.C., Maffett and Dowling,
Lieutenants Matson, Stitt, M.C., Aylmer, Hynes, Harrison, M.C.,
Sidwell, Chadwick, Condron and Utley.

Disembarkation was effected at Southampton on April 28th,
when the Battalion was at once railed to Bordon—where the 1st
Battalion was found to be also quartered—and was accommodated
in Quebec Barracks.

L

CHAPTER VIII

DISBANDMENT.

THE LAST SCENE.

THE process of disbandment was commenced so soon as the Battalion had taken up its quarters at Bordon, and the following table shows how this very trying measure was carried into execution.

There had, as already stated, been transferred to other corps prior to the departure of the Battalion from India, the following numbers and ranks :—

Warrant officers, 2 ; sergeants, 18 ; lance-sergeants, 4 ; corporals, 2 ; privates, 107 ; total, 133.

Further there were dispatched to their homes pending discharge :—

May 15th, 1922,	23	non-commissioned officers and men.
May 16th, 1922,	1	,, ,, ,, ,, ,,
May 22nd, 1922,	1	,, ,, ,, ,, ,,
May 23rd, 1922,	1	,, ,, ,, ,, ,,
May 26th, 1922,	3	,, ,, ,, ,, ,,
May 27th, 1922,	27	,, ,, ,, ,, ,,
May 30th, 1922,	29	,, ,, ,, ,, ,,
May 31st, 1922,	70	,, ,, ,, ,, ,,
June 1st, 1922,	58	,, ,, ,, ,, ,,
June 2nd, 1922,	58	,, ,, ,, ,, ,,
June 3rd, 1922,	58	,, ,, ,, ,, ,,
June 5th, 1922,	59	,, ,, ,, ,, ,,
June 6th, 1922,	61	,, ,, ,, ,, ,,
June 7th, 1922,	66	,, ,, ,, ,, ,,
June 8th, 1922,	46	,, ,, ,, ,, ,,
June 9th, 1922,	14	,, ,, ,, ,, ,,

ARRIVAL AT BORDON, 1922.

[Photo Gale & Polden, Ltd.

June 10th, 1922,   5 non-commissioned officers and men
June 16th, 1922,   1 ,,   ,,   ,,   ,,   ,,
June 21st, 1922,   3 ,,   ,,   ,,   ,,   ,,
June 23rd, 1922,   1 ,,   ,,   ,,   ,,   ,,
June 26th, 1922,   5 ,,   ,,   ,,   ,,   ,,
July 3rd, 1922,   2 ,,   ,,   ,,   ,,   ,,
July 5th, 1922,   2 ,,   ,,   ,,   ,,   ,,
July 7th, 1922,   4 ,,   ,,   ,,   ,,   ,,
July 11th, 1922,   1 ,,   ,,   ,,   ,,   ,,
July 14th, 1922,   36 ,,   ,,   ,,   ,,   ,,
July 17th, 1922,   5 ,,   ,,   ,,   ,,   ,,
July 21st, 1922,   5 ,,   ,,   ,,   ,,   ,,
July 28th, 1922,   6 ,,   ,,   ,,   ,,   ,,
July 31st, 1922,   6 ,,   ,,   ,,   ,,   ,,

      Total   ... 657

Transferred to other units, 30.

Discharged from hospital direct, 3.

Remaining in hospital, 1.

Remaining in India, 4.

The strength on March 1st, 1922, was therefore 828, including the 133 transferred before leaving India.

The following was the disposal of the commissioned ranks of the Battalion :—

Remained in India : Lieutenants C. M. Craig-McFeely, D.S.O., M.C., H. G. Harcourt, D.S.O., M.C., D. C. A. Shepard, T. E. Flewett, D. S. Norman and G. D. B. Russell.

Retired from the Army : Captains J. H. Allan and T. Brady, Lieutenants G. W. C. Bolster, A. G. L. Sidwell, M. A. Condron and E. G. Utley.

Posted to Home Units : Brevet Lieutenant-Colonel J. McD. Haskard, C.M.G., D.S.O., Major D. French, Captains C. G. Carruthers, M.C., G. W. B. Tarleton, M.C., and C. W. Maffett, Lieutenants H. B. Harrison and T. A. H. Chadwick.

Remaining for Disposal : Lieutenant-Colonel G. S. Higginson, Major S. G. Smithwick, O.B.E., Captains J. M. Mood, O.B.E.,

M.C , J. D. Glegg, M.C., and J. E. Dowling, Lieutenants C. Matson, M.C., W. H. Stitt, D.S.O., M.C., H. G. Aylmer, W. H. Hynes and M. H. FitzGerald.

Army Council Instruction No. 279 dated May 25th, 1922, contained full and final instructions as to the procedure to be followed in the disbandment of the South Irish regiments, and therein it was directed that battalions were as early as possible to be reduced to the following cadre, viz. : 1 commanding officer, 1 second in command, 1 president mess committee, 1 adjutant, 1 quartermaster, 1 regimental quartermaster-sergeant, 1 company sergeant-major, 1 company quartermaster-sergeant, 1 armourer (R.A.O.C.), 1 orderly room sergeant, 1 orderly room clerk, 1 sergeants' mess treasurer, 1 officers' mess sergeant, 1 transport sergeant, 4 sergeants, 8 corporals, 4 quartermaster storemen, 4 clerks, 5 batmen and 30 privates.

On May 30th it was notified that " His Majesty the King has honoured the Battalion by intimating his intention to receive the Colours of the Battalion and place them in safety in St. George's Hall, Windsor Castle. The handing-over ceremony will take place at 11.30 a.m. on June 12th."

On the same day there was published in Battalion Orders the following farewell letter from Major-General C. D. Cooper, C.B., the Colonel of the Royal Dublin Fusiliers, transmitting a kindly message of good-bye from Field-Marshal His Royal Highness the Duke of Connaught, Colonel-in-Chief :—

" *To* THE OFFICER COMMANDING,
     " 2/ROYAL DUBLIN FUSILIERS.

" FAREWELL ORDER TO THE 2ND ROYAL DUBLIN FUSILIERS, THE OLD TOUGHS.

" *It is with feelings of deep sorrow that I write these few lines of farewell to you all, and sign my name for the last time as your Colonel, which great honour was granted me on March 13th, 1910—over twelve years ago. I was appointed to the Old Toughs on July 8th, 1868, as an Ensign and I have always had the greatest pride in the Battalion and*

[*Photo, Gale & Polden, Ltd.*

COLOUR PARTIES OF "OLD TOUGHS" AND "BLUE CAPS" WHO PROCEEDED WITH THE COLOURS TO WINDSOR CASTLE TO HAND OVER SAME TO H.M. THE KING, 12TH JUNE, 1922.

"BLUE CAPS"			"OLD TOUGHS"	
Sergt. T. Doyle	Sergt. A. D. Connolly		C.Q.M.S. G. Sexton	C.Q.M.S. P. Keogh
Major J. P. Tredennick, D.S.O., O.B.E.	C.S.M. A. Cullen	Major T. J. Carroll-Leahy, D.S.O., M.C.	Capt. J. M. Mood, O.B.E., M.C.	C.Q.M.S. J. Jones	Capt. C. G. Carruthers, M.C.

in its splendid, and very old, traditions. I feel sure every one of you will always have a deep affection and pride in your old Corps.

" May every good luck attend you all from your ever good-wisher, comrade and admirer.

(*Signed*) " C. D. COOPER, *Major-General.*
" *Colonel, The Royal Dublin Fusiliers.*"

The letter from H.R.H. the Duke of Connaught was as follows :

" *To* MAJOR-GENERAL C. D. COOPER, C.B.,
    " COLONEL OF THE ROYAL DUBLIN FUSILIERS.

" *It is with a feeling of great sorrow that I bid farewell to the Royal Dublin Fusiliers. I have been your Colonel-in-Chief for nearly twenty years and have seen you in many parts of the Empire. The Regiment has been in existence over 250 years, and, as the Madras and Bombay Fusiliers, its history is practically the history of our Indian Empire. On its Colours are names such as Plassey, Arcot, Wandiwash, Seringapatam and Lucknow which show where the Regiment has fought.*

" *It is always a sign of great achievements when a regiment receives a nickname, and the ' Blue Caps ' extorted that title from the enemy in the great Indian Mutiny, and the ' Old Toughs ' earned their nickname by their tenacity in the Mahratta Wars.*

" *As the Royal Dublin Fusiliers the Regiment continued to play its part in the history of the British Empire. It earned the admiration of all at Talana Hill and in the battles for the relief of Ladysmith. In the late war it is only necessary to point to the landing at Gallipoli, and to Ypres, to show that the Regiment has never ceased to keep up the standard of its former achievements in India. It was to me a source of great pride when Her Majesty Queen Victoria appointed me Colonel-in-Chief of the Regiment. I have presented their Colours to both Battalions and, in the late war, I inspected battalions of the Regiment in France and Palestine.*

" *All ranks will share with me my satisfaction in knowing that His Majesty the King is going to take charge of our Colours, and it is my earnest hope that, in the future he may be able to restore its Colours to the Regiment, should the country again need its services.*

" In wishing good-bye to all ranks, I know that every man of the Royal Dublin Fusiliers will maintain his pride in the Regiment, and will never forget that he has worn its uniform and will ever help to maintain its glorious traditions.

    (Signed) " ARTHUR, Field-Marshal.

     " Colonel-in-Chief Royal Dublin Fusiliers."

To both these letters the Commanding Officer made reply ; to Major-General Cooper he wrote :—

    " QUEBEC BARRACKS,

     BORDON, HANTS,

      " 30*th* May, 1922.

" MY DEAR GENERAL,

 " We ' Old Toughs ' thank you most sincerely for your farewell order ; it is all the more appreciated coming from you who spent most of your long service in the Regiment as an ' Old Tough '.

 " Every rank has and does realise your constant interest in and love for the Regiment, and all the hard work you have performed for its welfare.

 " Your absence from our dinner and the reason for it are greatly deplored by all of us.

    " Very sincerely yours,

     (Signed) " G. S. HIGGINSON."

The following letter was sent by Colonel Higginson to the Colonel-in-Chief through Major-General Cooper :—

" SIR,

 " I have to acknowledge receipt of your letter of the 25th instant, enclosing farewell order of Field-Marshal His Royal Highness the Duke of Connaught, K.G., K.T., K.P., etc., Colonel-in-Chief.

 " Will you be good enough to convey the grateful thanks of all ' Old Toughs ' to His Royal Highness for his sympathy.

 " His Royal Highness's selection as our Colonel-in-Chief will always be a treasured recollection in our records.

COLOUR PARTIES ENTERING WINDSOR CASTLE,
JUNE 12TH, 1922.

" My Battalion respectfully asks that His Royal Highness will consent to convey to His Majesty the King our thanks and appreciation for his gracious honour in accepting our Colours for safe custody.

> *" I have the honour to be,*
> *" Sir,*
> *" Your obedient Servant,*
> (*Signed*) *" G. S. HIGGINSON, Lieutenant-Colonel,*
> *" Commanding The Old Toughs."*

Colonel Higginson's last order to the Battalion reads as under :—

" OLD TOUGHS !

" In saying good-bye to all of you, I am confident I am saying so to a body of men in whose keeping the traditions of our Battalion have never been in better hands.

" By your exemplary conduct and the fine spirit that abounds in all ranks and between all ranks, you are worthy exponents of our regimental motto.

" I thank you for the loyalty and unvaried support you have shown me in the all too short period in which I have had the honour to command you.   It has made that period a very pleasant one.

" Never forget the old Battalion and its customs and principles, for they will lead to success in whatever sphere your work may take you.

" Good luck to each one of you."

At a meeting of the Committee of Adjustment of the mess and other property of the Battalion it was decided that His Majesty the King be asked to accept a silver Centre-piece, and His Royal Highness the Duke of Connaught a silver Candelabra, as mementoes of the Battalion.   These offers were gladly and graciously accepted, and the gift of the Centre-piece was thus acknowledged, on behalf of His Majesty, by Lord Stamfordham, to Colonel Higginson :—

" The piece of Plate, which you were so kind as to bring here, has been handed to the King, and I am commanded to express to you and the officers of the Battalion under your command, His Majesty's sincere thanks for this gift, which will be treasured among the Royal

*Plate as a memento of a regiment, whose great and distinguished career
has been terminated through circumstances beyond its control."*

It had been decided that on Monday, June 12th, in St. George's
Hall, Windsor Castle, His Majesty the King should take over the
Colours of the undermentioned regiments :—

> The Royal Irish Regiment.
> The Connaught Rangers.
> The Prince of Wales's Leinster Regiment (Royal Canadians)
> The Royal Munster Fusiliers.
> The Royal Dublin Fusiliers.

While His Majesty had also agreed to receive a regimental engraving
from the South Irish Horse.

The detachments from these six corps travelled from Paddington
by the 9.55 a.m. train, arriving at Windsor at 10.42, the several
Colour parties then proceeding to the royal waiting room where
the Colours were unfurled, and the detachments were met by an
escorting party of 100 all ranks of the 3rd Battalion Grenadier
Guards and the Band of the Regiment.

The following officers and non-commissioned officers represented
the Royal Dublin Fusiliers :—

Colonel-in-Chief : F.M. H.R.H. the Duke of Connaught.
O.C. 1st Battalion : Lieutenant-Colonel C. N. Perreau, C.M.G.
King's Colour : Major J. P. Tredennick, D.S.O., O.B.E.
Regimental Colour : Major T. J. Carroll-Leahy, D.S.O., M.C.
Company Sergeant-Major A. Cullen, Sergeant T. Doyle,
Sergeant A. D. Connolly.
O.C. 2nd Battalion : Lieutenant-Colonel G. S. Higginson.
King's Colour : Captain J. M. Mood, O.B.E., M.C.
Regimental Colour : Captain C. G. Carruthers, M.C.
Colour-Sergeant J. A. Jones, Colour-Sergeant P. Kehoe, Colour-
Sergeant G. Sexton.

The Colour parties formed up in the following order in the
station yard :—Royal Irish Regiment, Connaught Rangers, South
Irish Horse, the Prince of Wales's Leinster Regiment, the Royal

[*Photo, "Daily Sketch."*

HANDING OVER THE COLOURS TO H.M. THE KING IN ST. GEORGE'S HALL,
WINDSOR CASTLE, JUNE 12TH, 1922.

Munster Fusiliers, and the Royal Dublin Fusiliers—and, preceded and followed by half the escort of the 3rd Battalion Grenadier Guards, marched up to the Castle, the Band playing the march-past of each regiment in turn.   At the gateway of the Castle the troops were met by Lieutenant-Colonel the Marquis of Cambridge, Governor and Constable of Windsor Castle, and on arrival in the Great Quadrangle the Band formed up in front of the Grand Entrance and played " Auld Lang Syne " as the detachments marched under the archway and passed, with their Colours, out of sight.

The Irish Regiments—all that was left of them—formed up in line in St. George's Hall, facing the windows, and received Their Majesties, the King and Queen, with a Royal Salute, and a ceremony which was of a private, indeed almost of an intimate or personal, character, then commenced.

Having closely inspected the line the King then spoke as follows to the representatives of his Irish Regiments :—

" We are here to-day in circumstances which cannot fail to strike a note of sadness in our hearts.   No regiment parts with its Colours without feelings of sorrow.   A knight in days gone by bore on his shield his coat-of-arms, tokens of valour and worth ; only to death did he surrender them.   Your Colours are the record of valorous deeds in war and of the glorious traditions thereby created.   You are called upon to part with them to-day for reasons beyond your control and resistance. By you and your predecessors these Colours have been reverenced and guarded as a sacred trust—which trust you now confide to me.

" As your King I am proud to accept this trust.   But I fully realise with what grief you relinquish these dearly-prized emblems ; and I pledge you my word that within these ancient and historic walls your Colours will be treasured, honoured and protected as hallowed memorials of the glorious deeds of brave and loyal regiments."

Their Majesties the King and Queen shook hands cordially and in regretful farewell with every member of the several Colour parties.   Then followed to each commanding officer a few words expressive of very real sympathy—and then finally came the personal touch, when the King, the Head of the Army and the Fount of Military Honour, handed to each of the commanding officers a letter of farewell addressed specially to all ranks of each regiment, recalling its past history, and expressing once more in warmest terms His Majesty's grateful appreciation of services rendered to King and Empire.

The Royal Letter to the " Old Toughs " is as follows :—

" *To* THE OFFICERS, WARRANT OFFICERS, NON-COMMISSIONED OFFICERS AND MEN OF THE ROYAL DUBLIN FUSILIERS.

" *It is with feelings of no ordinary sorrow that I address you for the last time ; for I know that I am taking leave not merely of a fine regiment, but of great memories and great traditions which hitherto have been kept alive and embodied in you.*

" *You are the oldest of the British garrison in India. Your Second Battalion dates back to the time when Queen Catherine of Braganza brought Bombay as part of her dowry to King Charles II. ; your First Battalion to still remoter days. Stringer Lawrence, the teacher of Robert Clive, won many a victory with you. Clive led you to Arcot and Plassey ; Eyre Coote to Wandewash ; Forde to Condore. Your history is the history of early British dominance in India, and you have shown abundantly that you could fight as sternly in South Africa and in Europe as in the East Indies.*

" *To me it is a very mournful task to bid you farewell—I have always taken the greatest pride in your past history, but if the glory of any fighting men be safe, then most assuredly safe is yours.*

" *You have your Colours, your trophies, and your household gods, which are dear to you as honour itself. You have thought fit to entrust your Colours to me for custody, and I am very proud to take charge of them, to be preserved and held in reverence at Windsor Castle as a perpetual record of your noble exploits in the field.*

" *Meanwhile be very sure that, with or without external monument, the fame of your great work can never die.*

" *I thank you for your good service to this Country and the Empire, and with a full heart I bid you—Farewell.*

(*Signed*) " GEORGE, R.I."

To the Officers, Warrant Officers, Non-Commissioned Officers and Men
of the Royal Dublin Fusiliers.

It is with feelings of no ordinary sorrow that I address you for the last time; for I know that I am taking leave not merely of a fine regiment, but of great memories and great traditions which hitherto have been kept alive and embodied in you.

You are the oldest of the British garrison in India. Your second battalion dates back to the time when Queen Catherine of Braganza brought Bombay as part of her dowry to King Charles II.; your first battalion to still remoter days. Stringer Lawrence, the teacher of Robert Clive, won many a victory with you. Clive led you to Arcot and Plassey; Eyre Coote to Wandewash; Forde to Condore. Your history is the history of early British dominance in India, and you have shown abundantly that you could fight as sternly in South Africa and in Europe as in the East Indies.

To me it is a very mournful task to bid you farewell—I have always taken the greatest pride in your past history, but if the glory of any fighting men be safe, then most assuredly safe is yours.

You have your Colours, your trophies and your household gods, which are dear to you as honour itself. You have thought fit to entrust your Colours to me for custody, and I am very proud to take charge of them, to be preserved and held in reverence at Windsor Castle as a perpetual record of your noble exploits in the field.

Meanwhile, be very sure that, with or without external monument, the fame of your great work can never die.

I thank you for your good service to this Country and the Empire, and with a full heart I bid you—Farewell.

George R.I.

12th June, 1922.

And so there, and in such manner, was the final act of disbandment accomplished, and thenceforth nothing remains of five splendid Irish regiments but those banners in St. George's Hall at Windsor Castle—" where memory sleeps."

Disbandment dragged its slow length along for a few days further, and then on July 31st it was curtly and officially announced that " the 2nd Battalion The Royal Dublin Fusiliers ceases to exist from this date," and so there passed from the British Army a Regiment which came into existence in the days of the Stuart Kings, and which, as the Bombay Regiment, the Bombay European Regiment, the 1st Bombay European Regiment, the 1st Bombay European Fuziliers, the 1st Bombay Fuziliers, the 103rd Royal Bombay Fusiliers, and as the 2nd Battalion The Royal Dublin Fusiliers, had fought in Asia, Africa and Europe on nearly every occasion with success, never without honour.

> Farewell the plumed troop and the big wars
> That make ambition virtue!   O, Farewell!
> Farewell the neighing steed and the shrill trump,
> The spirit-stirring drum, the ear-piercing fife,
> The royal banner and all quality,
> Pride, pomp, and circumstance of glorious war!
> And, O, you mortal engines, whose rude throats
> The immortal Jove's dread clamours counterfeit.
> Farewell! *

Othello, Act iii, Scene 3.

TO-MORROW we will turn the page and, lo !
　The book is done, as though the master-hand
That wrote it failed ; as though the tide had passed,
　And found the story only writ in sand.

Where will we hear the music that once throbbed,
　When bugles sang *réveillé*, and the roll
Of drums caught hard the breath of men ?　The notes
　Are silent as the passing of a soul.

Only in some cathedral aisle there droops
　Dead blazoning of regiments whose van
Still haunts the crumbling battlements of Spain,
　Or jagged passes of Afghanistan.

Graves scattered in the desert by the gloom
　Of the Crimean waters—will they stay
Unthought of when the casket that enshrined
　Their memory lies shattered in the clay ?

Dear perished names, struck from the roll, although
　No smirch was on them !　For a little space
They will be ours still ; then we, too, are gone,
　And none to tell where lies their resting-place.

Only among the graves the peasant folk
　Will linger on their names in Picardy
Or Flanders ; and the Turk in wonder tend
　Their hard-won tomb upon Gallipoli.

But in their island-home, whose name they bore
　Imperiously beyond the farthest main,
They are forgot : no tardy trumpet-call
　Can lure the vanished legions back again.

By kind permission of " Periscope."

"PLASSEY" "BUXAR"
"GALLIPOLI"
YPRES
"CARNATIC" "MYSORE"
SPECTAMUR AGENDO
THE ROYAL DUBLIN FUSILIERS
GREAT WAR 1914-1918
102ND "BLUE CAPS"
103RD "OLD TOUGHS"
THE ROYAL DUBLIN FUSILIERS
Raised ·1644·
Disbanded ·1922·
FIELD MARSHAL H.R.H. THE DUKE OF CONNAUGHT & STRATHEARN K.G. ETC
COLONEL-IN-CHIEF
Presented to :-
In recognition of his faithful service and as a memento of the old regiment on its disbandment.
Lt. Col
Commanding 2nd Bn The Royal Dublin Fusiliers. Old Toughs

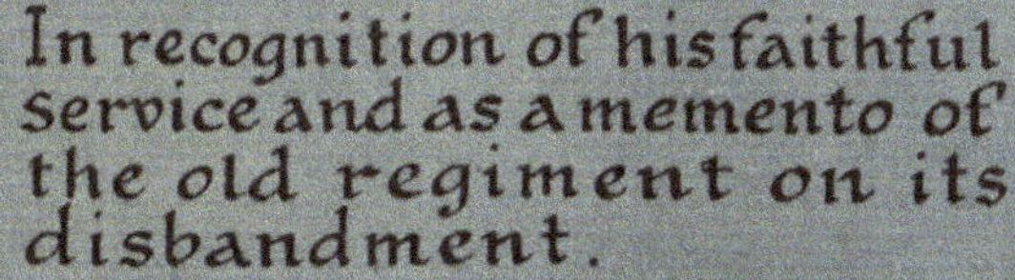

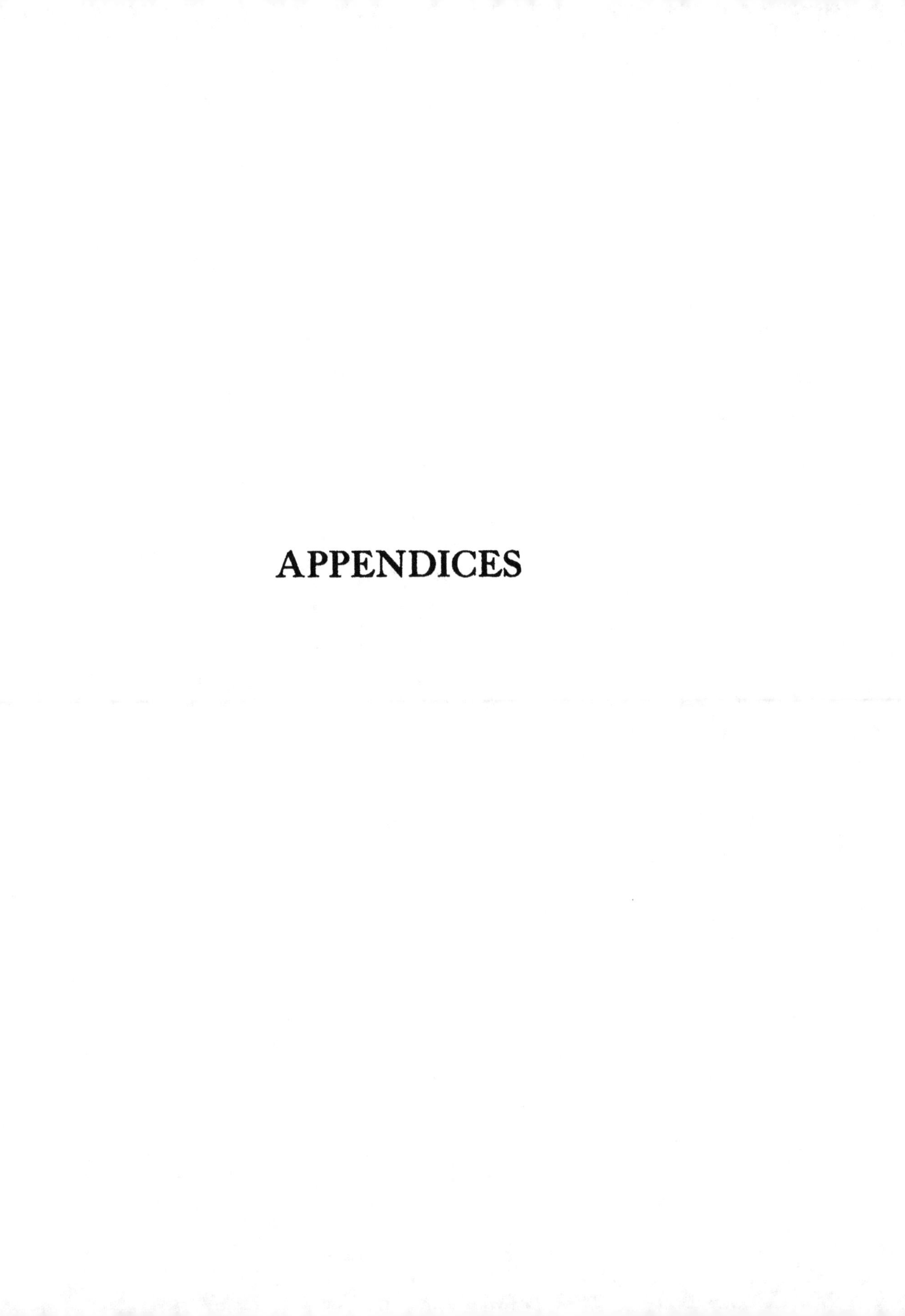

APPENDICES

[*Appendix One*

FIRST SEVEN DIVISIONS.

Commemoration, Albert Hall, December 15th, 1917.

Banner presented to Officers 2nd Battalion Royal Dublin Fusiliers by the following retired officers, relations, and friends of the Regiment, in ever Glorious Memory of all those gallant heroes who fell for their King and Country, and in deep appreciation of the conspicuous bravery and splendid conduct of all ranks. August to November, 1914.

Mrs. Gladys Alexander.
Miss A. M. Bacon.
Major and Mrs. Baker.
Mrs. Barry.
Col. and Mrs. Godfrey Bird.
Mr. W. E. Brown.
Capt. W. Bruce.
Miss E. M. Campbell.
Mrs. J. Campbell.
Miss Weyland Champion.
Col. Clark.
Mrs. Clover.
Major-Gen. and Mrs. C. D. Cooper.
Mrs. Dale.
Lieut.-Col. J. R. Dick.
Lieut.-Col. Dickinson.
Capt. Harris.
Lieut.-Col. and Mrs. M. J. Hickley.
Brig.-Gen. and Mrs. Tempest Hicks.
Mrs. D. M. Higginson.
Lieut.-Col. and Mrs. W. P. Holmes.
Lieut.-Col. C. A. Kerr.
Mrs. Percy Maclear.
Mrs. Mary Maclear.

Mrs. Duncan.
Mrs. James Duncan.
Mr. Edward Earl.
Brig.-Gen. F. P. English.
Miss Frankland.
Miss M. Frankland.
Mrs. H. French.
Miss M. J. Gilmore.
Lieut.-Col. and Mrs. F. W. Graham
Lieut.-Col. and Miss Godwin.
Lieut.-Gen. Sir Alex. and Lady Godley.
Dr. M. L. Grosvenor.
Major The Hon. and Mrs. Hobart-Hampden.
Mrs. Haskard.
Col. and Mrs. Pearse.
Lieut.-Col. Povah.
Dr. R. S. Purefoy.
Mrs. Ridley.
Col. and Mrs. Riddell.
Mrs. C. Rickards.
Mrs. Robinson.
Mrs. Rooth.
Col. H. S. Sheppard.
Mrs. Shewan.

Lieut.-Col. Bernard Maclear.
Col. and Mrs. R. H. Mansel.
Mrs. Dennis Murphy.
Major O'Connell.
Mrs. Ogilvy.
Mrs. O'Meara.
Mrs. O'Neill.
Mrs. W. H. O'Neill.
Mrs. Philby.
Mrs. Ralph Philby.
Mr. H. St. I. B. Philby.
Lieut. R. M. Philby, R.I.M.
Miss Marion Parsons.

Mrs. Charles Smith.
Major W. Seton.
Mr. Semmence.
Major and Mrs. Shaw.
Mrs. Supple.
Mr. R. Terrence.
Major F. L. C. Thomas.
Mr. E. Towell.
Mrs. Twist.
Capt. and Mrs. Vincent.
Mrs. Weldon.
Miss Margaret Wilson.

The following is taken from *The Times* of December 17th, 1917

AN IMPRESSIVE SCENE.

Khaki and black were almost the only colours to be seen in the audience that crowded the Albert Hall on Saturday afternoon for the choral commemoration of the First Seven Divisions. The khaki of the 700 officers, N.C.Os. and men who represented the survivors of that heroic army could be seen in the boxes and in the galleries above, in masses or patches up to the dizzy heights.

Black was everywhere ; for although this was a ceremony of proud commemoration, not of mourning, the civilian audience consisted almost entirely of the relatives of those gallant men, and of them few are not bereaved. Round the tiers hung—richly, not gaily—the Regimental and other banners, made for the occasion, embroidered in many instances by women who had travelled from the ends of the kingdom for the honour of putting in a few stitches. If feeling ran high, if at one portion of the ceremony the building rang with shouts and cheers and clapping of hands, the general tone was subdued, grave, and earnest.

There was not, probably, a single unoccupied place, nor even a foot of standing room in the highest gallery, when the King, the Queen, Queen Alexandra, Princess Mary, Princess Alice (Countess of Athlone), and Princess Victoria, with their suites, entered the

Royal boxes, the whole audience standing while the National Anthem was played.   Field-Marshal Lord French, who commanded the " contemptible little Army," was there.   Sir John Jellicoe and other naval officers occupied a box on the right of the Royal party, just beneath a great white ensign ; and among others to be seen in the assembly were Mr. Balfour, Lord Derby, Mr. Fisher (Minister of Education), Brigadier-General Turner, and Brigadier-General Gloster.   The programme (performed under the conductorship of Dr. Hugh P. Allen, by an orchestra led by Mr. William H. Reed, and a chorus composed of members of the Bach choir and members of the Royal Albert Hall Choral Society) consisted of English music only, and followed a scheme of ideas.   The first item, Sir Edward Elgar's " Cockaigne (in London Town)," stood for the light-hearted world before the war, Dr. Ralph Vaughan-Williams's song for chorus and orchestra, " Towards the Unknown Region " (Whitman's " Darest thou now, O Soul "), spoke the " challenge to the great adventure."   " The cost " was figured by Mr. Howell's " Elegy for Strings," composed in memory of a friend killed in this war ; " the achievement," by Mr. Arthur Somervell's Ode, " To the Vanguard, 1914," composed to the poem by Miss Beatrix Brice— " O little mighty Force that stood for England," which first appeared on the front page of *The Times*.   The solo in this was sung by Miss Lillian Stiles-Allen.   Next came " separation," expressed by a motet for unaccompanied voices, which is one of the " Songs of Farewell " written by Sir Hubert Parry for the years 1914-16 ; and Sir C. Villiers Stanford's song for bass solo and chorus (the vocalist being Mr. Plunket Greene), summed up, in the words of Sir Henry Newbolt's poem, " Farewell," the sacrifice, death, re-union, and immortality.

When the concert was over Mr. Balfour and Lord Derby came upon the platform, just in front of which stood a laurel-wreathed replica of Mr. Richard Belt's bust of Lord Kitchener, the original of which, cast from captured cannon, is in the War Office.   It was Mr. Balfour's office to read the verses of Ecclesiasticus (xliv., 1—14), beginning " Let us now praise famous men," which are familiar to

M

all public school and university men as the passage commonly read in commemoration of the founders and benefactors of colleges.

" There be of them, that have left a name behind them, that their praises might be reported.

" And some there be, which have no memorial; who are perished, as though they had never been; and are become as though they had never been born; and their children after them. . . .

" But these were merciful men, whose righteousness hath not been forgotten. . . .

" Their seed shall remain for ever, and their glory shall not be blotted out."

This commemoration—ringingly pronounced—of heroes, renowned or nameless, was followed in profound silence and reverence. When it was over, Lord Derby came forward to read the Order of Battle of the First Seven Divisions. First came the names of Field-Marshal Lord French, the principal officers of the General Staff, and the commanders of the four Army Corps; and loud were the cheers from the khaki-filled boxes and galleries at the names of the great captains. Then the trumpets and drums of the Coldstream Guards, under Major Mackenzie Rogan, broke into the " Fall In," and from the far end of the hall in marched the pipers and drums of the Scots Guards, to swing straight up the centre, playing as they went. Lord Derby then resumed his reading, taking Division by Division, giving the name of the officer commanding and the numbers and titles of the several units of Cavalry, Artillery, Engineers and Infantry in each. And when the " Regulars " had all been announced, Lord Derby read out the Yeomanry and the Territorial Battalions which had left as reinforcements to the Expeditionary Force before November 23rd, 1914. Cheering and clapping and crowing came from the gallant 700; cries of " Good old this ! " and " Good old that ! " and an interchange of professional banter.

From the liveliness of these moments the gathering passed to a different mood. The singing by all present of the hymn " For all

NIL DESPERANDVM
SPECTAMVR·AGENDO
WITTENBERG
GUSTROW
GIESSEN
DOEBERITZ
TORGAU
BAYREUTH
HEIDELBERG
SENNE
BING
KARLSRUHE
LANGENSALZA
MÜNSTER
MAGDEBURG
FORCHEIM
FRIEDRICHSFELD
ZERBST
LIMBERG
PHILIPSBURG

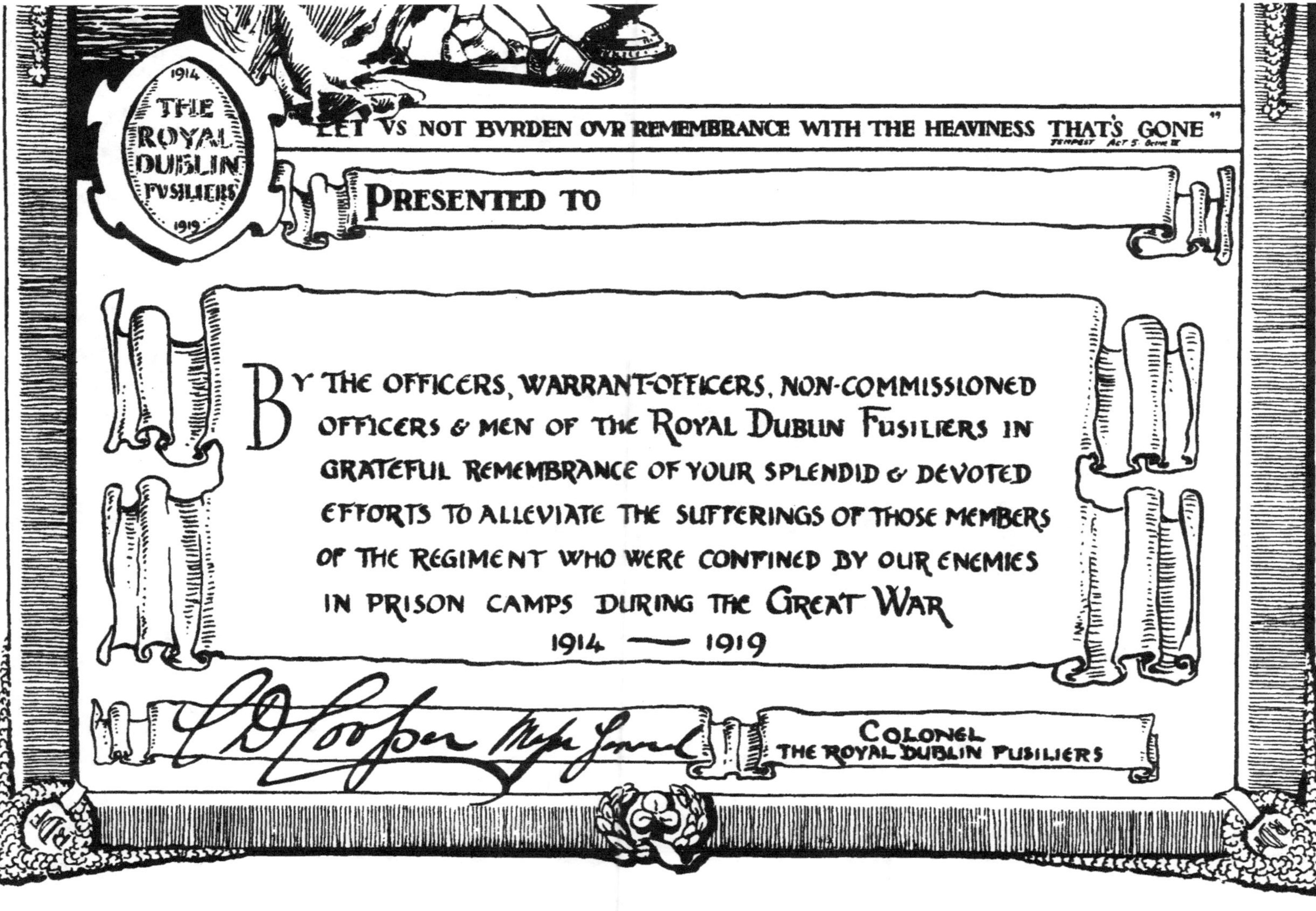

THE ROYAL DUBLIN FUSILIERS
1914 1919
"LET VS NOT BVRDEN OVR REMEMBRANCE WITH THE HEAVINESS THAT'S GONE"
TEMPEST ACT 5. SCENE II
PRESENTED TO
BY THE OFFICERS, WARRANT-OFFICERS, NON-COMMISSIONED OFFICERS & MEN OF THE ROYAL DUBLIN FUSILIERS IN GRATEFUL REMEMBRANCE OF YOUR SPLENDID & DEVOTED EFFORTS TO ALLEVIATE THE SUFFERINGS OF THOSE MEMBERS OF THE REGIMENT WHO WERE CONFINED BY OUR ENEMIES IN PRISON CAMPS DURING THE GREAT WAR
1914 — 1919
COLONEL
THE ROYAL DUBLIN FUSILIERS

the Saints," to Dr. Vaughan-Williams's music, left many eyes filled with tears, which were openly wiped away while the drums and trumpets challenged sorrow with the Réveillé. As their inspiriting clangour died down a voice was heard calling from somewhere below the Royal box for three cheers for the First Seven Divisions ; but the assembly was scarcely in the mood for cheering. The final singing of the National Anthem provided a more appropriate outlet for feeling that was as grave as it was tense.

The music, all by living Englishmen, was wisely chosen and splendidly sung. After the Cockaigne Overture, which gained point by being taken at a brisk pace, came the manliest of English choral works—Vaughan-Williams's " Towards the Unknown Region." It goes deep below the show of things to the great thoughts they embody, and in doing that it set the right tone of feeling for the commemoration. Its breadth and stature are realized for the first time in this large space, which, with such an audience and occasion, seemed its natural home. That could not be said of Howell's " Elegy " for strings which followed it, for it was with difficulty audible ; but we should have been sorry to miss hearing a work which so worthily, and with such promise, upholds the true tradition.

With Somervell's " To the Vanguard," now performed for the first time, the appeal was more direct and in so far less deep. The fault lay mainly with the poem, which has no impact, and contains no memorable line. Though the music does all that can be done with it, somehow it never sounded the note we were waiting to hear—that moment which lifts the particular to the universal, the transitory to the eternal. Miss Stiles-Allen made much of her short opportunity with a melodious soprano solo, and the tones of her voice, though tremulous, rang true.

The finest effort of the choir, consisting of the Bach Choir supported by the Royal Choral Society, was heard in Parry's " There is an old belief." The words offer a sound starting point, but the motet is a conspicuous instance of the power music has to illumine their message and drive home their meaning, as the setting sun and the incoming tide unite to transfigure a tall ship entering port. It was fitly followed by the song for bass and chorus, " Farewell," in which we did not know whether we were most deeply moved by Newbolt's words or Stanford's tones or Plunket Greene's voice, or by the time and place that made them what they were. Certainly Mr. Greene can seldom have had a more difficult task, or a task have shown more clearly what response the right man can make to it.

If there was anything in which the programme erred it was in a certain uniformity ; we wanted something like the Rondo of Beethoven's Funeral March sonata to cheer us up, and there would have been good military precedent for giving that turn to sad thoughts. Yet, perhaps, it was best as it was, now—and " For all the Saints," to Vaughan-Williams's tune, met the need in the best way. It would have been still better if the audience had noticed the intimation that their help was desired ; some day they will sing it, and will realize what it has rescued them from.

The orchestra owned some human imperfections, but atoned for them nobly. Mr. Darke's organ accompaniment was in good taste. Dr. Allen's conducting is a wonderful combination of energy, modesty, and good sense. No one would get the impression that he was " doing it all "—yet he does a great deal, or, more truly, has done it before entering the room. He is there merely as one among others labouring for a cause he has at heart.

The following telegram was sent on Saturday to the Commanders-in-Chief of the British Forces in France, Italy, Egypt, Salonika, and Mesopotamia :—

" Those assembled to-day in the Albert Hall, London, at a meeting honoured by the presence of their Majesties the King and Queen and Queen Alexandra to commemorate the heroic deeds of the First Seven Divisions, send their warmest greetings to all absent comrades and wish them good luck and a victorious homecoming."

OFFICERS' ROLL OF HONOUR

Bn.

4 Addis, Thomas Henry Liddon, Lt., k. in a., 21/3/18.
   Agnew, Andrew Eric Hamilton, Capt., died, 3/11/18.
11 Allen, Arthur Haviland, 2/Lt. (Tp.), k. in a., 4/10/17.
10 Allgood, George, 2/Lt. (Tp.), k. in a., 15/4/17.
1 Anderson, Denis Vipont Friend, Capt., k. in a., 25/4/15.
1 Andrews, William, 2/Lt., k. in a., 25/4/15.
10 Armstrong, Charles Martin, 2/Lt. (Tp.), k. in a., 8/2/17.
8 Bagley, Arthur Bracton, M.C., Capt., d. of w., 29/10/18.
3 Bankes, Edward Nugent, Capt., k. in a., 26/4/15.
11 Barre, Gerald Benedict, 2/Lt., killed, 9/8/18 (and R.A.F.).
10 Barrett, Hebron, Lt. (Tp.), d. of w., 27/3/18.
4 Bate, Alfred Francis, Lt., k. in a., 14/3/15 (att. Leinster Regt.).
4 Beddoes, Henry Roscoe, Lt.-Col., drowned, 15/1/19 (att. 9th Bn.).
8 Belas, Reginald Charles William, 2/Lt. (Tp.), k. in a., 21/3/18.
5 Bell, Alfred Roy Lancaster, 2/Lt., d. of w., 17/5/15.
7 Bell, Lee, 2/Lt. (Tp.), k. in a., 17/10/18.
1 Bernard, Robert, Lt., k. in a., 25/4/15.
11 Blackwell, Walter, Lt. (Tp.), d. of w., 28/9/18.
   Boles, Robert Stephen, 2/Lt. (Tp.), d. of w., 6/5/18.
2 Boulter, Jack Edward Hewitt, M.C., 2/Lt., d. of w., 15/10/18.
   Bourke, Bertram Walter, Capt., k. in a., 9/5/15.
1 Boustead, Lawrence Clive, Lt., k. in a., 28/6/15.
5 Boyd, Frederick Ennis, 2/Lt., k. in a., 20/5/17.
9 Boyd, William Hatchell, 2/Lt., k. in a., 9/9/16.
7 Boyle, John Kennedy, M.C., Lt., died, 21/10/18 (P. of W.).
11 Bradley, John McDonald, Lt. (Tp.), d. of w., 30/9/18.
10 Brereton-Barry, William Roche, 2/Lt., k. in a., 16/8/17.
6 Broun, Richard Clive McBryde, Lt. (Tp.), k. in a., 6/12/15.
8 Burns, Robert Henry, 2/Lt. (Tp.), k. in a., 1/11/18.
   Burroughs, Bernard Prendergast, 2/Lt., d. of w., 16/3/17.
11 Byrne, Edward Aloysius, 2/Lt. (Tp.), k. in a., 24/4/17.
8 Cahill, Thomas Laurence, 2/Lt. (Tp.), k. in a., 26/3/18.
9 Callear, Herbert, Capt. (Tp.), k. in a., 16/8/17.
9 Carrette, Albert Ernest, 2/Lt. (Tp.), k. in a., 27/4/16.
10 Carroll, Patrick, 2/Lt. (Tp.), k. in a., 8/2/17.
6 Carruth, John, Lt. (Tp.), d. of w., 10/10/18 (att. R. Ir. Rif.).
   Church, Frederick James, 2/Lt., k. in a., 10/5/15.
8 Clarke, George Alexander, 2/Lt. (Tp.), k. in a., 21/3/18.
5 Clarke, Wilfred John, 2/Lt., d. of w., 9/9/16.
6 Clery, Daniel Richard, Lt., k. in a., 10/8/15.
   Cliff, Harold Martin, Lt.-Col., died, 1/2/17.
10 Close, Charles Paul, 2/Lt. (Tp.), d. of w., 14/11/16.
   Close, Henry Burke, Lt., died, 1/11/18 (att. 1/2 Bn.).
7 Clover, Harwood Linay, Lt. (Tp.), died, 25/12/16 (and R.F.C.).
   Colles, Arthur Grove, Capt., k. in a., 12/3/15 (att. R. Ir. Rif.).
2 Conroy, Bernard, 2/Lt., k. in a., 5/7/15.
2 Considine, Christopher Daniel, 2/Lt., k. in a., 24/5/15.
7 Cooney, Charles Robert, 2/Lt. (Tp.), k. in a., 9/10/16 (att. R. Ir. Rif.).
9 Cooney, Edmund Luke, 2/Lt. (Tp.), k. in a., 4/6/17.
1 Corbet, Reginald Vincent Campbell, Lt., k. in a., 25/4/15.
   Cowley, George Evelyn, Major (Tp.), d. of w., 18/6/18 (P. of W.).
9 Coyne, John Joseph Aloysius, 2/Lt. (Tp.), k. in a., 10/8/17.

N

Bn.

1	Crawford, Sydney George, 2/Lt., drowned, 10/10/18.
7	Crichton, Aleck Godfrey, 2/Lt. (Tp.), k. in a., 16/8/15.
10	Cross, Henry Hazelock, 2/Lt. (Tp.), k. in a., 13/11/16.
	Cuffey, Maurice O'Connor, Lt., k. in a., 20/5/15.
7	Cunningham, Bernard Camelis Josh, Capt. (Tp.), k. in a., 21/3/18.
4	Cusack, Reginald Ernest, 2/Lt., died, 15/4/15.
5	Daly, Arthur Charles de Burgh, 2/Lt., k. in a., 9/9/16.
2	Damiano, Walter Henry Alexander, 2/Lt., d. of w., 2/7/16.
3	Davies, Charles Bernard, Lt., k. in a., 9/6/16.
8	Davies, Noel John, 2/Lt. (Tp.), k. in a., 27/4/16.
1	De Lusignan, Raymond, Lt., k. in a., 25/4/15.
3	Dillon, Edeveain Charles Barclay, 2/Lt., k. in a., 13/10/16.
6	Dinan, George Albert, 2/Lt. (Tp.), k. in a., 9/9/16.
8	Doherty, John, 2/Lt. (Tp.), k. in a., 16/8/17.
2	Donovan, Cyril Bernard, M.C., Lt. (Tp.), k. in a., 25/3/18.
7	Doran, Louis Godfrey, 2/Lt. (Tp.), k. in a., 23/10/16.
1	Dowling, Frederick, Lt., k. in a., 7/8/17.
2	Doyle, Christopher, 2/Lt., k. in a., 15/8/17.
6	Doyle, John Joseph, Lt. (Tp.), k. in a., 10/8/15.
8	Drury, William Symes, Lt. (Tp.), killed, 29/1/16.
1	Dunlop, George Malcolm, Capt., k. in a., 25/4/18.
2	Dunlop, John Gunning Moore, 2/Lt., k. in a., 27/8/14.
3	Dunne, John Geoffrey David Baird, Lt., died, 12/11/18.
4	Dunwoody, John Myles, 2/Lt., k. in a., 4/5/17.
4	Edwards, Brian Wallie, 2/Lt., died, 10/11/18 (and R.A.F.).
	Edwards, William Victor, Capt., k. in a., 29/12/17.
3	Ellis, Robert Percy, 2/Lt., d. of w., 6/4/18 (as P. of W.).
2	Falkiner, George Stride, 2/Lt., k. in a., 16/8/17.
4	Ferguson, James Ernest, 2/Lt., k. in a., 20/4/17.
1	Fetherstonhaugh, Edwyn, Major, d. of w., 27/4/15.
5	Finlay, Robert Alexander, Lt., k. in a., 9/5/15 (att. 1 R. Ir. Rif.).
10	Fitzgerald, Robert William, 2/Lt. (Tp.), d. of w., 4/10/18.
10	Fitzgibbon, G. J., Lt. (T./Capt.), k. in a., 20/11/17.
7	Fitzgibbon, Michael Joseph, Lt. (Tp.), k. in a., 15/8/15.
2	Floyd, Henry Murrell, Capt., k. in a., 28/6/15.
9	Forde, John Patrick, 2/Lt., d. of w., 16/8/17.
1	Frankland, Thomas Hugh Colville, Brevet-Major, k. in a., 25/4/15.
6	Freeney, Patrick Joseph, 2/Lt. (Tp.), k. in a., 8/10/18 (att. 198 L.T.M.B.).
2	French, Charles Stockley, Lt., k. in a., 25/4/15.
3	Gaffney, James, M.C., Capt. (Tp)., k. in a., 8/10/18.
	Gage, John, Capt., died, 7/11/16.
11	Gault, Arthur Alexander, 2/Lt. (Tp.), d. of w., 10/10/18.
	George, Herbert Duncan King, Lt., died, 6/4/17 (and R.F.C.) (P. of W.).
2	Gibson, Henry William, 2/Lt. (Tp.), d. of w., 27/11/16.
3	Girvin, Colin Bertram, Capt. (Tp.), died, 5/11/18.
9	Good, Thomas Henry, Capt. (Tp.), d. of w., 8/9/16.
1	Gradwell, George Francis, 2/Lt., k. in a., 28/2/17.
3	Graham, Cecil Hollingsworth, 2/Lt., k. in a., 19/9/16 (att. T.M.B.).
6	Graham, George Lyons, 2/Lt. (Tp.), k. in a., 17/8/17.
4	Gray, George, 2/Lt., killed, 28/4/16.
3	Gray, Meredith, 2/Lt., k. in a., 16/8/16 (att. 10 R. Ir. Rif.).
7	Greaves, Eric, M.C., Lt. (Tp.), d. of w., 21/11/18.
5	Green, Arthur Vivian, 2/Lt., k. in a., 17/8/17.
1	Green, Harold, 2/Lt. (Tp.), k. in a., 28/2/17.
1	Grimshaw, Cecil Thomas Wrigley, D.S.O., Major, k. in a., 26/4/15.
10	Guisani, St. John Joseph Vincent Anthony, 2/Lt. (Tp.), k. in a., 13/11/16.
11	Gyves, John James, 2/Lt. (Tp.), k. in a., 3/6/18.

Bn.

   Hackett, Henry Robert Theodore, 2/Lt., k. in a., 2/11/15.
3 Haigh, John Caleb, 2/Lt., k. in a., 2/10/18 (att. 1 R. Ir. Rif.).
   Halligan, Matthew, Lt., k. in a., 18/11/17 (and R.F.C.).
   Haines, Alec C., Lt., d. of w., 8/5/15.
   Hall, John Ramsay Fitz-Gibbon, 2/Lt., k. in a., 24/5/15.
8 Hamilton, Geoffrey Cecil Monck, 2/Lt. (Tp.), k. in a., 7/9/16.
4 Handyside, Thomas Fosbery, Lt., k. in a., 29/12/17.
6 Hare, Edward Henry, 2/Lt., k. in a., 23/9/17 (att. Yorks. Regt.).
7 Hare, George, Lt. (Tp.), k. in a., 27/12/17.
3 Harold-Barry, J., Capt., k. in a., 24/5/15.
11 Harty, Wilfrid, 2/Lt., k. in a., 8/8/17.
11 Harvey, John Alan, 2/Lt. (Tp.), k. in a., 20/11/17.
   Head, Henry d'Esterre, Lt., d. of w., 1/6/15.
4 Heenan, Thomas George Graudon, 2/Lt., k. in a., 21/3/18.
2 Helby, John Alfred Hasler, 2/Lt., d. of w., 3/8/16.
9 Hickey, Robert Francis, 2/Lt., d. of w., 16/8/17.
7 Hickman, Poole Henry, Capt. (Tp.), k. in a., 15/8/15.
1 Higginson, William Frederick, Capt., k. in a., 25/4/15.
11 Howden, Francis William, 2/Lt. (Tp.), d. of w., 30/3/18.
11 Howell, Reuben Harrison, 2/Lt. (Tp.), d. of w., 29/3/18.
8 Hughes, Bryan Desmond, M.C., Capt. (Tp.), k. in a., 6/8/18.
4 Humphrey, William, M.C., 2/Lt., d. of w., 24/10/18.
   Hunter, Ronald Gordon, 2/Lt., d. of w., 25/4/18 (P. of W.).
3 Inglis, Douglas Ian, 2/Lt., k. in a., 7/2/17.
2 Ingoldby, Roger Hugh, 2/Lt. (Tp.), k. in a., 1/7/16.
1 Jackson, William, 2/Lt. (Tp.), k. in a., 30/9/18 (att. 23 R. Fus.).
11 Jackson, Herbert, 2/Lt. (Tp.), k. in a., 21/3/18.
3 Johnson, Richard Digby, Major, k. in a., 24/5/15.
   Jones, Samuel Victor Charles, Lt., d. of w., 23/9/16.
3 Jones-Nowlan, Thomas Chamney, 2/Lt., d. of w., 27/5/17 (att. 1st Bn.).
   Judd, Frederick George Kerridge, 2/Lt. (Tp.), k. in a., 24/5/15.
7 Julian, Ernest Lawrence, Lt. (Tp.), d. of w., 8/8/15.
11 Karney, David Noel, T/2/Lt. (A./Capt.), k. in a., 21/3/18.
7 Kee, William, M.C., A./Capt., d. of w., 24/3/18 (as P. of W.).
11 Keenan, John, Capt. (Tp.), died, 26/3/17.
   Kempston, Robert James, Lt., k. in a., 24/5/15.
3 Kennedy, Arthur St. Clair, 2/Lt., died, 6/3/15.
9 Kettle, Thomas Michael, Lt. (Tp.), k. in a., 9/9/16.
5 Kidson, Charles Wilfrid, Lt., k. in a., 17/10/18.
2 Killingley, Hastings Grewatt, Lt., k. in a., 23/10/16.
   King, Robert Anderson Ferguson Smyly, 2/Lt., d. of w., 23/5/15.
2 Lemass, Herbert Justin, 2/Lt., k. in a., 23/10/16.
2 Le Mesurier, Frederick Neil, Capt., k. in a., 25/4/15.
2 Loveband, Arthur, C.M.G., Lt.-Col., k. in a., 25/5/15.
8 Lowe, Joseph, 2/Lt. (Tp.), k. in a., 26/3/18.
9 McAllister, Charles, A./Capt., k. in a., 27/5/18.
   McBrien, Hubert John, 2/Lt. (Tp.), k. in a., 4/11/18.
9 MacCarthy, Cornelius Aloysius, 2/Lt., drowned, 19/7/17.
   McCreery, Mona J. M., Capt., died, 21/10/18.
10 McCusker, Patrick Joseph, Lt. (Tp.), k. in a., 13/11/16.
3 MacDaniel, James, 2/Lt., k. in a., 18/8/17 (and R.F.C., 57 Sqd.).
6 McGarry, William Frederick Cecil, 2/Lt. (Tp.), k. in a., 10/8/15.
11 McGuinness, John Norman, 2/Lt., k. in a., 21/3/18 (att. 2 R. Mun. Fus.).
2 McGuire, Brian, 2/Lt., k. in a., 14/9/14.
7 Machutchison, William Frederick, Lt., k. in a., 26/3/18.
2 Maclear, Basil, Capt., k. in a., 24/5/15.
   Maclear, Percy, Major, k. in a., 30/8/14 (W. Africa).

Bn.

4 McLoughlin, James Patrick, Lt., d. of w., 24/5/15 (att. R. Ir. Rif.).
4 Macnamara, George Frederick, 2/Lt., k. in a., 17/8/16 (att. 8th Bn.).
2 Macnamara, Maccon John, 2/Lt., k. in a., 26/3/18.
5 Mcnulty, Michael John, Lt., k. in a., 4/9/15 (att. 9th Bn.).
11 Mallen, William James, 2/Lt. (Tp.), k. in a., 16/8/17.
9 Malone, Joseph James, 2/Lt. (Tp.), k. in a., 16/8/17.
10 Mansfield, Harold Barton, 2/Lt. (Tp.), k. in a., 13/11/16.
5 Marchant, Charles Stewart, 2/Lt., k. in a., 4/6/17.
8 Marlow, Charles Dwyer, 2/Lt. (Tp.), k. in a., 17/8/17.
6 Martin, Charles Andrew, Capt., k. in a., 6/12/15.
3 Martin, Geoffrey Clogstoun, 2/Lt., k. in a., 2/8/16.
3 Martin, Richard Archer Walcott, 2/Lt., k. in a., 16/8/17.
1 Maunsell, Edward Richard Lloyd, Capt., k. in a., 1/7/16.
8 Maxwell, Thomas, 2/Lt. (Tp.), k. in a., 9/9/16.
10 Mehegan, Daniel Joseph, 2/Lt., k. in a., 21/3/18.
11 Millar, James Roland, Capt. (Tp.), k. in a., 16/8/17.
8 Monson, William Herbert, Capt. (Tp.), d. of w., 7/9/16.
1 Mooney, Francis, 2/Lt., k. in a., 28/2/17.
9 Mooney, David George, 2/Lt. (Tp.), k. in a., 16/8/17.
1 Moore, Athelstan, D.S.O., Major (Bt.-Lt.-Col.), d. of w., 14/10/18.
5 Moran, Gerald Charles, Lt., d. of w., 26/5/15.
2 Morgan, John Walter Rees, 2/Lt., k. in a., 1/7/16.
6 Mortimer, William Lionel Gueritz, 2/Lt. (Tp.), d. of w., 10/8/15.
8 Murphy, Edward, 2/Lt. (Tp.), k. in a., 21/3/18.
5 Murphy, James Neville Herbert, 2/Lt., k. in a., 10/5/15.
9 Murphy, William Joseph, Capt. (Tp.), k. in a., 9/9/16.
10 Neilan, Gerald Aloysius, Lt. (Tp.), killed, 24/4/16.
6 Nesbitt, William Charles, 2/Lt. (Tp.), k. in a., 15/8/15.
1 Nolan, James, M.C., D.C.M., 2/Lt., d. of w., 29/9/18.
3 Nolan-Martin, Alfred John, Capt., d. of w., 22/2/17 (att. M.G.C.).
6 O'Carroll, Francis Brendon, 2/Lt. (Tp.), k. in a., 10/8/15.
1 O'Hara, Henry Desmond, D.S.O., Lt., d. of w., 29/8/15.
9 O'Kearney-White, Ernest Francis, 2/Lt., k. in a., 9/9/16.
5 O'Neill, Frederick, 2/Lt., k. in a., 13/11/16.
3 Palmer, David Adams, M.C., A./Capt., d. of w., 25/3/18 (att. Tank Corps).
10 Palmer, Samuel William, Lt., k. in a., 27/3/18.
  Pedlow, William, M.C., A./Capt., k. in a., 12/10/18.
3 Peel, Charles William, 2/Lt., k. in a., 24/4/15.
5 Perrier, Hargrave Carroll Lumley, 2/Lt., k. in a., 8/11/18.
4 Persse, Dudley Eyre, Lt., d. of w., 1/2/15.
1 Philby, Denis Duncan, Lt., k. in a., 12/11/14 (att. R. Munster Fus.).
7 Pige-Leschallas, Gilbert, Capt. (Tp.), k. in a., 15/8/15.
9 Pither, Harold Francis, 2/Lt., k. in a., 6/7/16.
9 Potter, Robert John, 2/Lt. (Tp.), k. in a., 16/8/17.
8 Poulter, Henry Chapman, Capt. (Tp.), d. of w., 29/11/17.
5 Powell, Frederick William, 2/Lt., died, 20/1/17.
6 Preston, A. J. D., Capt., k. in a., 15/8/15.
  Prince-Smith, Donald St. Patrick, Lt. (Tp.), k. in a., 24/10/17 (att. R.F.C.,
[16 Sqd.).
11 Quigley, Christopher, 2/Lt. (Tp.), k. in a., 21/3/18.
11 Quinn, John Patrick, Lt. (Tp.), d. of w., 20/6/17.
  Ransome, Frederick Ronald, 2/Lt., k. in a., 27/5/18.
3 Reavie, Wilfred Laurance, 2/Lt., k. in a., 16/8/17.
3 Reid, Bernard, 2/Lt. (Tp.), k. in a., 28/6/16.
9 Richards, Leslie John, 2/Lt. (Tp.), k. in a., 1/8/17.
6 Richards, William Reeves, Capt. (Tp.), k. in a., 15/8/15.
  Ridley, Herbert Leslie, M.C., Lt. (A./Capt.), k. in a., 15/7/17.

Bn.

 1 Roberts, William John, 2/Lt., k. in a., 21/3/18.
   Robertson, Eric Hume, 2/Lt., k. in a., 21/3/18 (att. 48 T.M.B.).
11 Rogers, James Joseph, 2/Lt. (Tp.), k. in a., 28/3/18.
 1 Rooth, Richard Alexander, Lt.-Col., k. in a., 25/4/15.
 4 Rose-Cleland, Alfred Middleton Blackwood Bingham, Lt., k. in a., 1/7/16.
 7 Russell, Alexander James, Lt. (Tp)., k. in a., 15/8/15.
10 Russell, Thomas Wallace, 2/Lt. (Tp.), k. in a., 13/11/16.
 4 Saffery, Leslie Hall, 2/Lt., k. in a., 1/7/16.
 4 Salvesen, Edward Maxwell, 2/Lt., k. in a., 25/4/15.
 8 Sheridan, Leonard, Capt., k. in a., 26/3/18.
 8 Sheridan, Richard Brinsley, Lt. (Tp.), killed, 7/3/16.
 2 Shine, James Owen Williams, Capt., k. in a., 16/8/17.
   Sparrow, Francis, 2/Lt., k. in a., 25/4/15.
 6 Stanford, Donovan Edward, 2/Lt., k. in a., 21/3/18.
 6 Stanton, Robert, 2/Lt., k. in a., 7/8/15.
 4 Stewart, Joseph, 2/Lt., k. in a., 16/8/17.
   Storrar, Andrew Wynne, 2/Lt., k. in a., 16/8/17 (att. 48 T.M.B.).
 7 Sutherland, William, M.M., 2/Lt. (Tp.), k. in a., 7/10/18.
 2 Sutton, Robert William, 2/Lt., T./Capt., k. in a., 16/10/15.
   Taylor, Adrian Aubrey Charles, Capt., k. in a., 28/6/15 (att. Egypt Police),
                                     (killed serving with 1st Bn. R.D.F.).
   Taylor, John Arthur Harold, 2/Lt., k. in a., 24/9/15.
 3 Thomas, Daniel Gwyn, 2/Lt., k. in a., 25/5/15.
   Thompson, Gerald Pittis Newman, Lt., k. in a., 4/5/18 (att. 8th Bn.).
 7 Tippet, Charles Henry, T./Major, k. in a., 7/8/15.
 7 Tobin, Richard Patrick, Capt. (Tp.), k. in a., 15/8/15.
10 Traverse, James Hector, 2/Lt., (Tp.), k. in a., 30/11/17.
11 Tumilty, Austin, 2/Lt., (Tp.), died, 10/11/17.
 3 Tweedy, Cecil Mahon, Lt., k. in a., 28/2/17.
 4 Tyndall, Joseph Charles, Lt., k. in a., 2/3/15 (att. R. Ir. Rif.).
 9 Tyner, Thomas Goodwin, 2/Lt. (Tp.), k. in a., 9/9/16.
 8 Valentine, Robert Lepper, Lt. (Tp.), d. of w., 30/4/16.
 5 Vardon, Evelyn Francis Claude, 2/Lt., k. in a., 10/5/16.
 9 Verley, Albert Stuart Leonard, T./Lt., k. in a., 16/8/17.
 9 Vigors, Arthur Cecil, 2/Lt. (Tp.), k. in a., 9/9/16 (att. R. Munster Fus.).
   Walkey, Francis Ashton, 2/Lt., k. in a., 17/10/18.
 2 Walsh, Lionel Percy, Capt. (Tp. Major), d. of w., 4/7/16.
 8 Walsh, Phillip James, T./2/Lt., k. in a., 30/11/17.
11 Ward, Bernard, 2/Lt. (Tp.), k. in a., 20/11/17.
 4 Warner, Douglas Redston, 2/Lt., k. in a., 1/7/16.
 7 Weatherill, Edward Theaker, 2/Lt. (Tp.), k. in a., 15/8/15.
 4 White, William, 2/Lt., k. in a., 25/4/15.
 3 Whitehead, Walter, 2/Lt., k. in a., 6/9/16.
11 Williams, John, 2/Lt. (Tp.), drowned, 22/10/17.
10 Wilson, Alexander Stewart, 2/Lt. (Tp.), k. in a., 20/4/17.
   Wilson, Denis Erskine, Major (Tp.), d. of w., 24/9/16.
   Wylie, J. R., 2/Lt., killed, 23/4/18 (att. R.A.F.).
   Young, Mervyn Cyril Nicholas Radford, 2/Lt., d. of w., 25/5/15.

Total number of Officers (*a*) Killed in Action, (*b*) Died of Wounds,
(*c*) Killed, other than in action, (*d*) Died, from natural causes, etc. :—

269.

OTHER RANKS ROLL OF HONOUR

Total number of Warrant Officers, Non-commissioned Officers and Men (*a*) Killed in Action, (*b*) Died of Wounds, (*c*) Killed, other than in action, (*d*) Died, from natural causes, etc. :—

4508.

A complete Roll of Soldiers of The Royal Dublin Fusiliers who were killed in action or died as the result of wounds, etc., can be found in " Part 73," The Royal Dublin Fusiliers, published by H.M. Stationery Office, at Imperial House, Kingsway, London, W.C.2, or 23, Forth Street, Edinburgh, or from Eason & Sons, Ltd., 41 & 42, Lower Sackville Street, Dublin.   Price 2/6.

List of
HONOURS AND REWARDS
issued to
THE ROYAL DUBLIN FUSILIERS
for services in
THE GREAT WAR, 1914-18.

NOTE.—*This list is as accurate as possible from data available. Omissions or errors may have occurred. It is hoped that notification of same will be sent to the publishers with a view to corrections being made should a second impression become necessary.*

VICTORIA CROSS.

Curtis, Horace Augustus ; sergeant (Regtl. No. 14107) ; 2nd Battalion. For gallantry in France. Invested 8/3/19. (*London Gazette*, 6/1/19.)

Downie, Robert ; sergeant (Regtl. No. 11213) ; 2nd Battalion. For gallantry in France. V.C. presented by the King at Sandringham, 8/1/17. (*London Gazette*, 25/11/16.)

Ockenden, James ; sergeant (Regtl. No. 10605) ; 1st Battalion. For gallantry east of Langemaarke, France. Invested 5/12/17. (*London Gazette*, 8/11/17.)

ORDERS OF BATH, St. MICHAEL AND St. GEORGE, DISTINGUISHED SERVICE ORDERS, MILITARY CROSSES, AND PROMOTIONS.

NAME.	RANK.	BATTALION.	HONOURS AND REWARDS.	"LONDON GAZETTE."
Alexander, Edward	2nd-Lieut.	1st	M.C.	24/9/18
Amos, Stephen Lewis	Lieut. (Temp. Capt.)	5th (attd. 152nd T.M.B.)	M.C.	4/6/17
Arnold, Norman Alfred	Temp. Lieut.	Attd. 1st	M.C.	11/1/19
Ashton, Frederic Ellis	Major (Temp. Lieut.-Col. York and Lanc. Regt.)	Attd. 7th	D.S.O.	3/6/18
Bagley, Arthur Bracton	Capt.	1st	M.C.	17/4/17
Bailey, Charles	Temp. Capt.	10th	M.C.	1/1/18
Barrett, Edward Robert	Temp. Lieut.	Attd. 2nd Bn. R. Innis. Fusiliers	M.C.	15/2/19
Beaumont, William Victor	2nd-Lieut.	Spec. Res. (attd. 2nd)	M.C.	17/9/17
Bellingham, Edward Henry Charles Patrick, Sir	Temp. Lieut.-Col. (Temp. Brig.-Gen.)	8th (R. of O., late Royal Scots)	D.S.O.	20/10/16
			C.M.G.	1/1/18
			Bt. Lieut.-Col. (A.O. 193)	30/1/20
Bergin, Charles Joseph	2nd-Lieut.	3rd (attd. 1st Bn. Royal Munster Fusiliers)	M.C.	2/12/18
Black, Walter Ian	2nd-Lieut.	Spec. Res. (att. 2nd)	M.C.	26/7/17
			Bar to M.C.	18/10/17
Boulter, Jack Edward Hewitt	2nd-Lieut.	2nd	M.C.	15/2/19
Brown, Frederic Elliott	Lieut.	3rd, and No. 84 Sqdn. R.F.C.	M.C.	22/4/18
			Bar to M.C.	22/6/18
Browne, Bernard Joseph (M.M.)	Temp. 2nd-Lieut.	11th (attd. 10th)	M.C.	18/2/18
Burke, John	Qr.Mr. and Hon. Maj.	2nd	M.C.	1/1/17
			D.S.O.	3/6/19
Burke-Savage, Ivan	Temp. Lieut.	10th (attd. 1st)	M.C.	11/1/19
Burrowes, Alfred Edward	Temp. 2nd-Lieut.	9th (attd. 2nd)	M.C.	1/1/18
Byrne, Louis Campbell	Lieut. (A./Capt.)	Attd. 2nd	M.C.	18/10/17
			Bar to M.C.	18/2/18
			D.S.O.	15/2/19
Byrne, Richard	Qr.Mr. and Hon. Lieut.	Service	M.C.	2/2/16
Calwell, William Maunsell	Temp. Capt.	10th (attd. 19th Entr. Bn.)	M.C.	16/9/18
Cameron, William Bedford St. George	Lieut.	3rd (attd. 1st)	M.C.	1/1/19
Carew, Robert John Henry	Capt.	—	M.C.	1/1/17
			Bt. Major	3/6/19
Carroll-Leahy, Thomas Joseph	Capt. and Bt. Major	—	M.C.	1/1/15
			Bt. Major	4/6/17
			D.S.O.	1/1/19

Name	Rank	Unit	Honour	Date
Carruthers, Colin Gordon	Capt.	—	M.C.	2/5/16
Carson, William Roland	Temp. Lieut.	Attd. 24th	M.C.	8/3/19
Chandler, George Hamilton	Lieut. (A./Capt.)	Spec. Res. (attd. 1st)	M.C.	26/11/17
Clarke, Norman Percy	Major	—	Bt. Lieut.-Col.	1/1/19
Clarke, Thomas Scott	Temp. 2nd-Lieut., 2nd Bn. R. Irish Regt.	Attd. 1st Bn.	M.C.	8/3/19
Corballis, Edward Roux Littledale	Capt.	R.D.F. and R.F.C.	D.S.O.	1/1/18
Cory, George Norton (D.S.O.)	Major	R.D.F.	Bt. Lieut.-Col.	18/2/15
			Bt. Col.	1/1/17
			C.B.	3/6/18
			Major-Gen.	1/1/19
Cox, John Francis	Temp. 2nd-Lieut.	10th	M.C.	26/1/17
Crawford, Gilbert Malcolm	Temp. Capt.	7th (attd. 2nd)	M.C.	8/3/19
Crozier, Herbert Charles	Capt.	1st	M.C.	3/7/15
Cullen, Patrick John	2nd-Lieut. (A./Capt.)	1st (attd. 1st Bn. R. Irish Rifles)	M.C.	1/2/19
Cuninghame, Henry Maurice Benedict Gun.	Lieut. (Temp. Capt.)	Spec. Res. (attd. 1st)	M.C.	17/4/17
Dalton, James Emmet	Temp. 2nd-Lieut.	7th (attd. 9th)	M.C.	20/10/16
Darling, Sydney George	Temp. 2nd-Lieut.	9th (attd. 1st)	M.C.	16/9/18
Davies, Douglas Joseph	2nd-Lieut.	5th (attd. 10th)	M.C.	3/6/18
Devlin, David	2nd-Lieut.	3rd (attd. 2nd)	M.C.	8/3/19
Devoy, Joseph Charles Brendin	Lieut. (A./Capt.)	1st	M.C.	4/2/18
Dickie, James MacNeece	Lieut.	—	M.C.	23/6/15
Dickie, T. W.	Capt.	Spec. Res.	Bt. Major	1/1/18
Dickson, Thomas Cedric Harold	Capt. (A./Major)	4th (attd. 13th Bn. R. Innis. Fus.)	M.C.	3/6/19
Dobbs, John Fritz Kivas	Capt.	—	M.C.	30/1/20
Dolan, John Joseph	2nd-Lieut.	Attd. 1st Bn. R. Ir. Rif.	M.C.	16/9/18
Donovan, Cyril Bernard	Temp. 2nd-Lieut (A./Capt.)	2nd	M.C.	1/1/18
Duff-Taylor, Squire	Capt.	4th (attd. 2nd)	M.C.	3/6/18
Egan, Nicholas Joseph	Temp. 2nd-Lieut.	9th	M.C.	27/7/16
Esmonde, James	Lieut.	Attd. 6th	M.C.	1/1/18
Fisher, Clarence George Cary	Temp. 2nd-Lieut.	8th (attd. 1st)	M.C.	16/9/18
Frankland, Thomas Hugh Colville	Capt.	—	Bt. Major	18/2/15
Franklin, Hubert Charles	Temp. 2nd-Lieut.	8/9th	M.C.	4/2/18
Gaffney, James	Lieut.	Spec. Res. (attd. 2nd)	M.C.	11/12/16
Galbraith, Hugh	Temp. 2nd-Lieut.	10th (attd. 19th Entr. Bn.)	M.C.	16/9/18

NAME.	RANK.	BATTALION.	HONOURS AND REWARDS.	"LONDON GAZETTE."
Gehrke, Richard Arthur	Lieut.	5th Spec. Res. (attd. 30th M.G. Coy.)	D.S.O.	25/11/16
Gibbs, George Howard	Lieut. (Temp. Major)	4th (attd. Vth Corps as T.M. officer)	M.C.	3/6/19
Glegg, James Donie	Capt.	2nd	M.C.	4/6/17
Goodbody, Jerome Marcus	Temp. Lieut.	10th (attd. 4th Bn. North'd Fus.)	M.C.	24/9/18
Greaves, Eric	Temp. Lieut.	7th (attd. 2nd)	M.C.	2/4/19
Grove, James Robert Wood	Capt. (Temp. Major)	—	Bt. Major	3/6/19
Guilbride, Francis Langford	Temp. Capt.	8th	M.C.	24/6/16
Hamlet, Francis Aylmer	Lieut. (A./Capt.)	3rd (attd. 2nd Bn. Tank Corps)	M.C.	3/6/19
Hannin, William Francis	Temp. Lieut.	6th	M.C.	8/3/19
Harcourt, Harry Gladwyn	Temp. Lieut. (A./Capt.)	Attd. 51st Bn. M.G.C.	M.C.	3/6/18
			D.S.O.	26/7/18
			Bar to D.S.O.	21/1/20
Harrison, Herbert Berkeley	Lieut. (A./Capt.)	1st (attd. 1/6th Bn. Glouc. R.)	M.C.	2/4/19
Harvey, William James	Lieut.	Spec. Res. and No. 102 Sqdn. R.F.C.	M.C.	22/6/18
Haskard, John McDougall	Major (Temp. Lieut.-Col.)	—	D.S.O.	1/1/17
			Bt. Lieut.-Col.	1/1/18
			C.M.G.	1/1/19
Hatt, Frederick (Regtl. No. 5531)	Regtl. Sergt.-Major	2nd	M.C.	1/1/17
			Bar to M.C.	16/9/18
Hawtrey, Eric Edmund Harcourt	Temp. 2nd-Lieut.	Attd. 1st	M.C.	8/3/19
Hayes, Hugh Joseph	Capt.	4th (attd. 6th)	M.C.	8/3/19
Healy, Christopher Francis	Temp. 2nd-Lieut.	8th	M.C.	1/1/17
Healy, Maurice	Capt.	4th (attd. Labour Corps)	M.C.	3/6/19
Heffernan, James Gerald Patrick	Temp. Capt.	9th	M.C.	4/6/17
			D.S.O.	3/6/19
Henchy, Alphonso Watson	Temp. 2nd Lieut.	10th	M.C.	24/1/17
Henderson, E. (Regtl. No. 10819)	Company Sergt.-Major	—	M.C.	18/2/15
Higgins, Richard Leo	Temp. 2nd-Lieut.	10th (attd. 9th Bn. R. Innis. Fus.)	M.C.	17/4/17
			Bar to M.C.	15/2/19
Higginson, Harold Whitla	Major and Bt. Col. (Temp. Lt.-Col., Temp. Brig.-Gen., Temp. Major-Gen.)	—	D.S.O.	14/1/16
			Bt. Lt.-Col.	3/6/16
			Bt. Col.	1/1/18
			Bar to D.S.O.	16/9/18
			C.B.	1/1/19

Name	Rank	Unit	Honour	Date
Hill, Charles	Temp. 2nd-Lieut.	9th	M.C.	17/9/17
Horrell, Joseph Betts	2nd-Lieut.	5th (attd. 2nd)	M.C.	8/3/19
Hughes, Bryan Desmond	Temp. 2nd-Lieut.	8th	M.C.	24/6/16
Hughes, James Allan	Lieut.	Spec. Res. (attd. 6th Bn. L. N. Lancs. R.)	M.C.	15/10/18
Humphrey, William	2nd-Lieut.	4th (attd. 2nd)	M.C.	15/2/19
Hunt, John Patrick	Temp. Capt. (now Temp. Major, A./Lieut.-Col.)	8th (now Gen. List)	D.S.O.	20/10/16
			Bar to D.S.O.	26/7/18
			C.M.G.	3/6/19
Hurst, Nicholas	Temp. 2nd-Lieut.	9th	M.C.	20/10/16
Hurst, Nicholas	Capt. 9th Gurkha Rif. (I.A.)	—	Bar to M.C.	10/6/21 to date 1/8/20
Jeffreys, Richard Griffith Bassett	Major (Temp. Lieut.-Col.)	2nd	D.S.O.	1/1/18
Jeffries, William Francis	Capt. (A./Major)	Spec. Res.	D.S.O.	26/7/18
Jowett, Harold Arthur	Temp. 2nd-Lieut.	1st (attd. 6th)	M.C.	25/11/16
Julian, Arthur Wellesley	Lieut.	Spec. Res. (attd. 10th)	M.C.	4/6/17
Kane, Michael Harry Kirkpatrick	Lieut.	Spec. Res. (attd. 10th)	M.C.	18/2/18
Kee, William	Temp. Lieut.	7th	M.C.	25/11/16
			Bar to M.C.	16/9/18
Kelly, Daniel	Temp. Capt. R.A.M.C.	Attd. 1st Bn.	M.C.	4/6/17
Kendrick, Edward Holt	Bt. Major (Temp. Lieut.-Col.)	R.D.F. (comdg. 11th Bn. Suffolk Regt.)	Bt. Major	4/6/17
			D.S.O.	1/1/18
			Bt. Lieut.-Col.	3/6/19
Kidd, William Ruddock	2nd-Lieut.	Spec. Res. (attd. 8th)	M.C.	22/9/16
Kiernan, Farrell Michael	Temp. Capt.	2nd	M.C.	2/4/19
Laffan, Paul	2nd-Lieut.	1st	M.C.	26/11/17
Lambkin, Douglas Raymond	Lieut.	5th (attd. 2nd)	M.C.	8/3/19
Law, Robert	Capt.	1st	M.C.	22/9/16
Lind, Walter Peterson	Temp. 2nd-Lieut.	8th	M.C.	24/6/16
Lindsay, Walter Charles (M.V.O.)	Lieut.-Col.	Spec. Res.	Bt. Col.	1/1/19
Little, Edward Gerald	2nd-Lieut.	Attd. 7th	M.C.	16/8/17
Lloyd, Percy Gamaliel Whitelocke	Temp. 2nd-Lieut.	11th (attd. 2nd)	M.C.	17/12/17
Loveband, Arthur	Lieut.-Col.	R.D.F.	C.M.G.	18/2/15
McElnay, George Herbert	Lieut.	3rd (attd. 2nd)	M.C.	8/3/19
McFeely, Cecil Michael	Lieut.	1st	M.C.	17/4/17
			D.S.O.	26/11/17
			Bar to M.C.	8/3/19

NAME.	RANK.	BATTALION.	HONOURS AND REWARDS.	"LONDON GAZETTE."
McGowan, Patrick John	Temp. 2nd-Lieut.	11th (attd. 1st)	M.C.	24/9/18
McGrath, John Augustine	Temp. 2nd-Lieut.	11th (attd. 9th)	M.C.	18/—/17
McKenna, John	2nd-Lieut.	3rd (attd. R. Ir. Rif.; attd. 107th L.T.M.B.)	M.C.	15/2/19
Mackenzie, Ian Dawson	2nd-Lieut. (Temp. Lieut.)	3rd (attd. 1st)	M.C.	1/1/18
McMahon, Vincent Matthew	Temp. 2nd-Lieut.	10th	M.C.	26/1/17
Maguire, Robert	Temp. 2nd-Lieut.	7th	M.C.	26/9/17
Massy-Westropp, Ralph Frederick Hugh	Capt.	R.D.F.	M.C.	3/6/19
Matson, Colin	Lieut.	Attd. 10th M.G.C.	M.C.	4/6/17
Matthews, Horace Lionel	Temp. Capt.	8th	M.C.	4/6/17
May, John Paul (M.M.)	2nd-Lieut.	Attd. 1st Bn. R. Muns. Fus.	M.C.	2/12/18
Miller, John Stephen	Temp. 2nd-Lieut.	Attd. 6th	M.C.	8/3/19
Molesworth, Edward Algernon	Major	—	D.S.O.	15/10/15
Monson, William Herbert	Temp. Capt	8th	M.C.	24/6/16
Montgomery, Malcolm Ronald	2nd-Lieut.	2nd	M.C.	18/2/18
Moore, A.	Major	—	Bt. Lieut.-Col.	1/1/18
Morris, Samuel	2nd-Lieut.	2nd	M.C.	15/2/19
			Bar to M.C.	2/4/19
Noblett, George Harris	2nd-Lieut. (A./Capt.)	3rd (attd. 1st)	M.C.	11/1/19
			Bar to M.C.	15/2/19
Nolan, James (D.C.M.)	2nd-Lieut.	1st	M.C.	16/9/18
O'Connor, Matthew John	2nd-Lieut.	R.D.F. and 50th Bn. M.G.C.	M.C.	15/2/19
O'Connor, Peter John	Temp. 2nd-Lieut.	8th	M.C.	24/6/16
O'Donnell, Michael Francis	Temp. 2nd-Lieut.	8th (attd. 1st)	M.C.	11/1/19
			Bar to M.C.	8/3/19
			2nd Bar to M.C.	8/3/19
O'Dowda, James Wilton	Lieut.-Col. (Temp. Brig.-Gen.)	R.D.F. and Staff	Bt.-Col.	3/6/16
			C.M.G.	22/12/16
			C.B.	7/2/18
O'Hara, Henry Desmond	Lieut.	—	D.S.O.	3/7/15
O'Malley, Victor David	Capt. (Temp.)	10th (attd. 2nd Bn. R. Muns. Fus.)	M.C.	18/6/17
			Bar to M.C.	15/2/19
O'Sullivan, Gerald Patrick	Lieut.	3rd (attd. 2nd)	M.C.	8/3/19
Oulton, William Plato	Capt.	Spec. Res. (attd. 1st)	M.C.	7/12/15
			Bar to M.C.	18/2/18

Name	Rank	Battalion	Honour	Date
Palmer, David Adams	2nd-Lieut.	Spec. Res. (attd. 8th)	M.C.	20/12/16
Parker, Harry (Regtl. No. 27362)	Coy. Sergt.-Major	8th	M.C.	14/11/16
Perreau, Charles Noël	Major (Temp. Col. 5/2/15, Temp. Brig.-Gen. Canadian Militia 10/12/18)	2nd (employed Colonial Office as Comdt. R.M.C. of Canada)	C.M.G. (for services in connection with the War)	4/6/18
Patterson, Ronald More	Lieut. (A./Capt.)	Spec. Res. (attd. 2nd)	M.C.	19/8/17
Pedlow, William	Lieut. (A./Capt.)	2nd	M.C.	18/2/18
Petit, Gorth	2nd-Lieut.	2nd	M.C.	18/2/18
Philippe, Douglas George	2nd-Lieut.	4th	M.C.	17/9/17
Popham, F. S.	Capt.	Spec. Res.	Bt. Major	1/1/18
Power, Frederick Thomas Alfred	Lieut.	2nd	M.C.	17/9/17
Quigley, Eugene Patrick	Temp. Lieut.	9th	M.C.	20/10/16
Renny, Lewis Frederick	Major	Staff, R.D.F.	D.S.O.	23/6/15
			Bt. Lieut.-Col.	1/1/17
			C.M.G.	1/1/18
			Bt.-Col.	3/6/19
Ridley, Herbert Leslie	Lieut. (Temp. Capt.)	1st	M.C.	1/1/17
Robinson, John Poole Bowring	Major	—	D.S.O.	4/6/17
			C.M.G.	1/1/19
Romer, Cecil Francis	Major and Bt. Lieut.-Col. (Temp. Col. on the Staff)	—	C.B.	18/2/15
			Bt. Col. and A.D.C. (extra) to the King	14/1/16
			C.M.G.	1/1/17
			Major-Gen.	1/1/19
Ross, Frederick George	2nd-Lieut.	5th (attd. 1st)	M.C.	11/1/19
Seymour, Evelyn Francis Edward	Major (A./Lieut.-Col.)	Attd. 10th	D.S.O.	1/1/18
Shadforth, Harold Anthony	Capt.	Attd. 6th	M.C.	8/3/19
Staples, John Vincent	Temp. 2nd-Lieut.	Attd. 2nd	M.C.	8/3/19
Stirke, Henry Richard	Temp. Major	9th (attd. 2nd)	D.S.O.	18/2/18
Stirling, Walter Francis	Major	R. of O. (late R.D.F.)	M.C.	1/1/18
			Bar to D.S.O.	8/3/19
Synnott, Frederick William	Capt.	Spec. Res. (attd. 10th and 19th Entr. Bn.)	M.C.	16/9/18
Thompson, Albert Charles	Temp. Lieut.-Col.	8th	D.S.O.	1/1/18
Thompson, Frank Septimus	Temp. Lieut.	8th	M.C.	24/6/16
Tippet, Herbert Charles Coningsby	Capt.	4th	M.C.	1/1/19
Treacher, Frederick	Lieut.	—	M.C.	14/1/16
			Bt. Major	3/6/19

NAME.	RANK.	BATTALION.	HONOURS AND REWARDS.	"LONDON GAZETTE."
Tredennick, James Paumier	Major (Temp. Lieut.-Col.)	—	D.S.O.	14/1/16
Watson, Ronald Macgregor	Capt.	—	D.S.O.	18/2/15
			Bt. Major	1/1/17
Weir, Andrew Herbert	Temp. Lieut.	Attd. 1st	M.C.	8/3/19
Weldon, Kenneth Charles	Major	—	D.S.O.	1/1/17
			Bt. Lieut.-Col.	3/6/19
White, Herbert	Temp. Lieut. (A./Capt.)	9th	M.C.	26/7/18
			D.S.O.	3/6/19
Whyte, William Henry	Lieut. (Temp. Major, A./Lieut.-Col.)	R. of O., R.D.F. (attd. 6th Bn.)	D.S.O.	1/1/18
Wolfe, William Cooper	2nd-Lieut.	Spec. Res. (attd. 2nd)	M.C.	16/9/18

THE ORDERS OF THE BRITISH EMPIRE, MILITARY DIVISION, 1914–20.

NAME.	RANK.	BATTALION.	HONOURS AND REWARDS.	"LONDON GAZETTE."
Bell, Lee	2nd-Lieut.	11th	M.B.E.	7/6/18
			Transferred to Mil. Div.	15/4/19
Bromilow, Walter	Col.	R.D.F.	C.B.E., Grade III	3/6/19
Byrne, Richard, M.C.	Capt. and Qr.Mr.	R.D.F.	O.B.E., Grade IV	3/6/19
De Courcey Wheeler, Samuel Gerald	Temp. Lieut.-Col.	R.D.F.	O.B.E., Grade IV, with effect from 3/6/19	12/12/19
Elsworthy, Alexander Lockhart	Capt.	2nd	M.B.E., Grade V	3/6/19
Halligan, Joseph Thomas	Capt.	R.D.F.	O.B.E., Grade IV	1/1/19
Henry, Hugh	Major (Spec. Res.)	Late 3rd	O.B.E., Grade IV, with effect from 3/6/19	12/12/19
Keegan, Michael, M.M.	Capt. (A./Major)	—	O.B.E., Grade IV	3/6/19
Lanigan-O'Keefe, François Stephen	—	R.D.F.	M.B.E., Grade V	3/6/19
Meldon, James Austen	Lieut.-Col.	4th	C.B.E., Grade III	3/6/19
Mood, John Muspratt, M.C.	Capt. (Temp. Major)	R.D.F., seconded M.G.C.	O.B.E., Grade IV	3/6/19
Moore, James Stuart Hamilton	T./Capt. (Gen. List, G.S.O.3, Intelligence H.Q., Eastern Command)	R.D.F.	O.B.E., Grade IV	3/6/19

NAME.	RANK.	BATTALION.	HONOURS AND REWARDS.	" LONDON GAZETTE."
Robinson, Robert Hervey StClair	Lieut.-Col.	5th	M.B.E., Grade V (transferred to Mil. Div., *Lond. Gaz.*, 23/1/20)	8/1/19
Seymour, Evelyn Francis Edward, D.S.O.	Major	R.D.F.	O.B.E., Grade IV	3/6/19
Shadforth, Harold Anthony	Capt.	R.D.F. (attd. 6th)	O.B.E., Grade IV	20/3/19
Shore, Alfred George	Lieut. (Temp. Capt.)	3rd	M.B.E., Grade V	3/6/19
Smithwick, Standish George	Major	R.D.F.	O.B.E., Grade IV	3/6/19
Tredennick, James Paumier, D.S.O.	Major	R.D.F. ; D.A.A.G. 63rd (R.N.) Div.	O.B.E., Grade IV	3/6/19
Ward, John	Major	R.D.F.	O.B.E., Grade IV	1/1/19

DISTINGUISHED CONDUCT MEDAL.

NAME.	REGTL. NO.	RANK.	BATTALION.	"LONDON GAZETTE."
Ainger, A. G....	28934	Private	6th	12/3/19
Alexander, E.	9529	Sergeant	8th	20/12/16
Benson, R. H., M.M.	27691	Sergeant	1st	12/3/19
Brophy, P.	4907	Sergeant	5th	24/1/17
Byrne, L.	8901	C.Q.M.S.	1st	22/10/17
Byrne, M.	16105	Private	9th	20/10/16
Byrne, S.	10774	A./Sergeant	1st	22/1/16
Callaghan, E.	17988	Private	2nd	15/3/16
Carrick, J.	16980	C.S.M.	9th	20/10/16
Chittenden, B.	7574	Bandsman	2nd	17/12/14
Conneys, P.	22813	L./Corporal	2nd	18/2/18
Cooke, W.	8672	A./Sergeant	2nd	30/6/15
Cooney, C.	10256	Sergeant	1st	6/9/15
Cowell, J.	14627	Sergeant	9th	20/10/16
Cullen, L.	6847	Sergeant	9th	19/11/17
Cullen, T.	10159	Sergeant	1st	1/1/19
Cullen, T.	10113	Private	1st	5/8/15
Cummins, W.	6603	C.S.M.	2nd	16/5/16
Bar to D.C.M.	—		—	12/3/19
Curley, W. P.	9508	Corporal	2nd	14/1/16
Delaney, P.	9364	Sergeant	1st	26/11/17
Dennis, J. M.	28923	Sergeant	6th	12/3/19
Devoy, J.	10335	Sergeant	1st	2/2/16
Doherty, J.	14507	C.S.M.	8th	24/6/16
Donfield, J.	10414	A./Corporal	9th	25/8/17
Dunne, D.	24580	Private	5th	24/1/17
Dyke, C. P.	9761	Private	1st	16/5/16
Ferguson, S.	6128	Sergeant	1st	3/7/15
Ford, J.	17811	Private	1st	16/11/15
Fox, H.	4823	C.S.M.	12th	30/1/20
Furphy, A.	18364	C.S.M. 2nd Bn. R. Mun. Fus.	Late 10th	3/10/18
Gaynor, F. J.	10090	Sergeant	2nd	18/2/19
Gibson, G.	16531	Corporal	1st	12/3/19
Gormley, T.	21233	Corporal	1st	5/12/18
Green, W. C. C.	27780	Sergeant	2nd	18/2/19
Bar to D.C.M.	—	—	—	18/2/19
Greenwood, L. B.	10544	C.S.M.	1st	12/3/19
Guest, A.	14153	A./Sgt.-Major	7th	2/2/16
Hall, R. S.	5039	C.S.M.	2nd	1/4/15
Halloran, M.	18341	Corporal	1st	3/6/19
Healy, R.	14705	Sergeant	8th	24/6/16
Hurley, J. F.	10374	Private	1st	1/1/17
Jennings, T.	10641	Private	1st	16/5/16
Kane, Patrick	10592	Corporal	1st	4/3/18
Kavanagh, R.	8087	Private	2nd	3/6/15
Kelly, C.	11330	Private	2nd	14/1/16
Kelly, E.	28182	Private	8th	1/1/18
Knight, H.	10515	Sgt. (A./R.S.M.)	2nd	1/1/19
Knightley, A. G.	24078	Sergeant	2nd	12/3/19
Lamb, P.	25439	Private	1st	5/12/18
Lennon, M.	15729	Sergeant	8th	24/6/16
Lowe, W.	29318	Private	10th	3/6/18
McManus, J., M.M.	40313	Sergeant	1st	12/3/19
McNamara, F.	10132	Corporal	1st	6/9/15
McPartlin, M., M.M.	21379	Sergeant	1st	30/10/18
Maloney, J.	7100	C.S.M.	2nd	14/1/16
Mangan, P. J.	16540	Sgt. (A./C.S.M.)	—	20/10/16
Moran, P.	7094	Private	2nd	3/6/18
Mulligan, F.	5834	Private	1st	26/11/17

NAME.	REGTL. NO.	RANK.	BATTALION.	"LONDON GAZETTE."
Murphy, H.	13190	A./R.S.M.	1st	22/10/17
Murray, F.	11371	Private	2nd	3/9/18
O'Brien, D.	20155	Sergeant	6th	12/3/19
O'Brien, J.	9581	L./Corporal	1st	26/11/17
O'Brien, J. F.	13886	C.S.M.	9th	19/11/17
O'Connor, J.	8746	Sergeant	6th	3/6/16
O'Keefe, J.	18763	Sergeant	8th	24/6/16
O'Leary, J.	10310	Sergeant	1st	3/6/19
Perrott, H.	9266	Sergeant	1st	12/3/19
Perry, F.	15834	Sergeant	2nd	18/2/19
Bar to D.C.M.	—	—	—	18/2/19
Roache, J.	9559	L./Sergeant	8th	20/10/16
Robinson, H.	14275	C.S.M.	7th	25/11/16
Ryan, E.	8669	Sergeant	2nd	14/1/16
Shanahan, M.	43052	A./Corporal	2nd	1/5/18
Smith, A.	8222	Sergeant	9th	20/10/16
Stafford, J.	27322	Private	Formerly 10th	3/10/18
Starkie, J.	15828	Private	20th	18/2/19
Stead, T. R.	17748	Private	1st	2/2/16
Stokes, J.	9150	A./Corporal	2nd	19/11/17
Tait, T.	14613	C.S.M.	8th	25/8/17
Waine, P.	11167	Sergeant	1st	22/10/17
Bar to D.C.M.	—	—	—	26/11/17
Wall, C. J.	24478	Corporal	1st	26/11/17
Watts, M.	9121	Private	1st	3/6/19

MILITARY MEDAL.

NAME.	REGTL. NO.	RANK.	BATTALION.	"LONDON GAZETTE."
Acheson, J.	7/13817	Sergeant	7th	22/1/17
Archbold, M.	27247	Private	9th	16/11/16
Aylward, J. J.	26464	L./Corporal	10th	19/2/17
Banbury, D.	27463	Private	1st	14/5/19
Barry, J.	12007	Private	6th	17/6/19
Bass, C. A.	24093	Sergeant	2nd	17/6/19
Bastin, W. G.	15244	Sergeant	9th	19/11/17
Batchelor, S.	22172	Pte. (A./Cpl.)	1st	21/10/18
Bates, H.	22204	L./Corporal	1st	11/11/16
Batley, B.	266607	L./Corporal	1/6th	18/6/17
Behan, T.	16451	Pte. (Cpl.)	6th	23/7/19
Benson, R. H.	27691	Corporal	1st	17/6/19
Billman, G.	27820	Private	6th	20/8/19
Bingham, J.	21139	Private	8th	10/8/16
Blades, J.	29697	Sergeant	2nd	17/6/19
Blake, W.	5837	Private	1st	17/4/17
Blayney, G.	16907	Private	2nd	19/11/17
Blostein, P.	43172	Private	2nd	17/9/17
Blyth, R. J.	13857	Sergeant	1st	11/2/19
Boland, F. C.	14920	L./Corporal	2nd	17/6/19
Boon, W.	10464	C.Q.M.S.	2nd	11/11/16
Bowen, C.	7/24250	Sergeant	7th	22/1/17
Boyce, P.	24274	Private	2nd	17/6/19
Bramble, D.	10882	A./Corporal	1st	17/4/17
Brannan, P.	22434	Private	6th	23/7/19
Brazington, J.	28936	Corporal	6th	17/6/19
Brennan, T.	5610	Private	2nd	11/11/16
Brierley, E.	20041	Private	8th	19/11/17
Brooks, J.	5008	Sergeant	1st	14/1/18
Brophy, P.	4907	Sergeant	1st	18/10/17
Brown, D.	21193	L./Corporal	8th	16/11/16

NAME.	REGTL. NO.	RANK.	BATTALION.	"LONDON GAZETTE."
Brown, G.	8471	Corporal	1st	17/6/19
Brown, T.	30741	Private	1st	11/2/19
Browne, B. J.	25579	Private	10th	19/2/17
Browne, T.	11515	Private	2nd	13/3/18
Brunton, G.	27907	Private	1st	21/10/18
Bryan, E.	13197	Sergt.-Dmr.	6th	9/7/17
Bryan, S.	43610	Private	6th	20/8/19
Buggy, M.	16108	Private	2nd	23/7/19
Burke, J.	10809	C.Q.M.S.	2nd	6/8/18
Burke, J.	8/17090	Sergeant	8th	14/9/16
Butler, C.	5696	Private	8/9th	23/2/18
Butler, T.	18631	Private	1st	14/1/18
Byrd, B.	16517	Sergeant	8th	16/11/16
Byrne, A.	14874	Sergeant	2nd	19/11/17
Byrne, J.	8868	Private	2nd	13/3/18
Byrne, J.	9590	Private	2nd	22/1/17
Byrne, J.	24807	Private	1st	11/2/19
Byrne, L.	8901	Sergeant	1st	11/11/16
Byrne, P.	1823	Private	1st	11/2/19
Byrne, T.	5460	Private	2nd	6/8/18
Cahill, J.	9829	Sergeant	2nd	29/8/18
Callaghan, J.	12699	Private	6th	17/6/19
Campbell, J.	8/15473	Sergeant	8th	14/9/16
Cantillon, R.	18613	Private	2nd	11/11/16
Carr, J.	1/17528	Private	7th	25/4/16
Carroll, J.	8249	Sergeant	2nd	6/1/17
Carroll, M.	9808	Private	1st	21/9/16
Carroll, P.	19392	Pte. (Cpl.)	9th	19/11/17
Carter, J.	8337	Private	2nd	4/2/18
Cashel, J.	43539	Private	2nd	14/5/19
Cassidy, F.	25659	Private	8th	16/11/16
Chambers, A. S.	24110	Private	2nd	23/7/19
Chisolm, A.	30570	Private	2nd	23/7/19
Christian, T.	20014	Pte. (L./Cpl.)	1st	29/8/18
Clark, J.	18369	Private	2nd	11/11/16
Cleary, W. P.	20103	Private	6th	9/12/16
Clifton, R.	14444	Private	8th	16/11/16
Clifton, W.	28444	Pte. (A./L./Cpl.)	2nd	23/7/19
Coen, T.	27995	Private	6th	17/6/19
Collier, R.	5452	Private	2nd	11/11/16
Connell, L.	29912	A./Sergeant	1st	11/2/19
Connolly, E. P.	30170	Private	1st	17/6/19
Connolly, P.	10237	Private	2nd	20/8/19
Conroy, P.	30520	Private	1st	17/6/19
Cooney, M.	26734	Private	9th	18/10/17
Crawford, T.	15544	L./Corporal	9th	17/9/17
Crealy, C.	6889	Sergeant	1st	14/1/18
Crosby, T.	15590	Private	9th	18/10/17
Crowley, W.	15794	Private	8th	10/8/16
Crudden, P.	10834	Corporal	2nd	19/11/17
Bar to M.M.	—	—	—	13/3/18
Cullen, M.	43524	Private	1st	17/6/19
Cummins, P.	25455	Private	1st	11/2/19
Cummins, W.	20881	Private	1st	11/2/19
Curran, P.	40549	L./Corporal	9th	17/9/17
Currie, A.	13039	Sergeant	1st	17/4/17
Dagg, P.	18621	Corporal	8th	17/9/17
Daintree, T.	43182	Private	2nd	13/3/18
Daly, P.	11669	Private	2nd	18/7/17
Bar to M.M.	—	—	—	13/3/18
2nd Bar to M.M.	—	—	—	14/5/19

NAME.	REGTL. NO.	RANK.	BATTALION.	"LONDON GAZETTE."
Davies, W.	265370	Private	6th	18/7/17
Davis, A. W.	43232	Private	2nd	19/11/17
Davis, F.	27357	Private	2nd	13/3/18
Delaney, J.	1/7874	Corporal	6th	14/1/18
Delaney, P.	9364	C.S.M.	1st	11/2/19
Devine, P.	12687	Corporal	6th	30/1/20
Devitt, W. H.	25570	Private	1st	17/6/19
Devlin, T.	22358	Pte. (L./Cpl.)	1st	14/5/19
Doherty, J.	18370	A./Corporal	2nd	11/11/16
Doherty, T.	27069	Private	9th	17/9/17
Donfield, J.	1/10414	Corporal	9th	22/1/17
Donohoe, T.	19582	Private	1st	14/1/18
Downie, R.	11213	Sergeant	2nd	11/11/16
Doyle, C.	19412	Cpl. (A./Sergt.)	10th	13/3/18
Doyle, J.	30818	Private	2nd	23/7/19
Doyle, P.	4998	Private	1st	14/1/18
Doyle, T.	31346	Corporal	1st	11/2/19
Drake, T.	10314	L./Corporal	2nd	14/5/19
Dunn, H.	28132	Private	2nd	13/3/18
Dunne, J.	21859	Private	2nd	13/3/18
Durr, T. F.	26956	A./Sergeant	2nd	18/7/17
Ebbitt, J.	15797	Sergeant	8th	17/9/17
Egan, F.	9771	Corporal	2nd	11/11/16
Bar to M.M.	—	Sergeant	—	17/6/19
Eglington, J. J.	10640	Sergeant	1st	17/6/19
Bar to M.M.	—	—	—	17/6/19
Ellis, J.	23063	Pte. (L./Cpl.)	8th	11/5/17
Elsey, F.	29011	Private	2nd	17/6/19
Fallon, J.	20102	Sergeant	1st	17/6/19
Felton, W.	16888	L./Corporal	8th	10/8/16
Finerty, J.	18472	L./Corporal	1st	18/10/17
Finlay, A.	10099	Corporal	1st	21/10/18
Finnie, J. R.	29250	Corporal	1st	14/5/19
Flynn, W.	20825	L./Corporal	1st	18/10/17
Foley, J.	40297	Private	1st (attd. 86th T.M. Bty.)	14/1/18
Ford, H. J.	15841	Corporal	2nd	17/6/19
Foster, R.	16223	Private	8th	17/9/17
Bar to M.M.	—	—	—	21/10/18
Foster, W.	40819	Private	1st	11/2/19
Fox, C.	9851	Private	1st (attd. 86th T.M. Bty.)	14/1/18
Fox, E. J. W.	15712	L./Sergeant	1st	6/8/18
Fox, J.	26503	Private	1st	11/2/19
Fray, W. A.	29150	Corporal	2nd	17/6/19
Freeman, P.	9223	Private	2nd	14/5/19
Fullerton, J.	9504	L./Corporal	2nd	11/11/16
Gable, E.	14757	Private	7th	6/1/17
Gaffney, M.	26965	Sergeant	2nd	21/12/16
Gallagher, J.	9536	Private	1st	9/7/17
Galvin, P.	19730	Private	6th	17/6/19
Gannon, P.	6/11864	Private	6th	22/1/17
Gasteen, F. E.	27317	Sergeant	1st	17/6/19
Geddes, R.	17577	Pte. (L./Cpl.)	1st	25/4/18
Gee, J.	40208	Private	1st	14/5/19
Gibson, G.	16531	Corporal	1st	11/2/19
Gill, A.	28816	Private	2nd	14/5/19
Gill, H. P.	40890	Pte. (L./Cpl.)	8th	19/11/17
Gleeson, R.	11481	Corporal	9th (attd. 48th T.M. Bty.)	17/9/17
Glynn, C.	8259	Pte. (L./Cpl.)	2nd	13/3/18

NAME.		REGTL. NO.		RANK.		BATTALION.			"LONDON GAZETTE."
Glynn, T. P.	...	23253	...	Corporal	...	1st	...	...	17/6/19
Bar to M.M.	...	—		—		—			17/6/19
Godden, C.	...	14315	...	Private	...	1st	...	...	29/8/18
Goldsborough, J.	...	30339	...	Pte. (L./Cpl.)	...	1st	...	...	11/2/19
Goode, A. E.	...	40535	...	Corporal	...	1st	...	...	11/2/19
Gorman, P.	...	11642	...	Private	...	2nd	...	...	19/11/17
Gorry, J.	...	24825	...	Cpl. (L./Sgt.)	...	1st	...	...	29/8/18
Goulding, M.	...	18944	...	L./Corporal	...	9th	...	...	21/8/17
Grace, L.	...	9692	...	Private	...	1st	...	...	14/1/18
Gray, R.	...	14245	...	Private	...	2nd	...	...	23/7/19
Green, M.	...	28739	...	Private	...	1st	...	...	18/10/17
Green, P.	...	23438	...	Private	...	2nd	...	...	13/3/18
Griffen, J.	...	8639	...	L./Corporal	...	9th	...	...	17/9/17
Guinane, T.	...	19342	...	Sergeant	...	1st	...	...	11/2/18
Halleron, J.	...	7075570	...	L./Corporal	...	2nd	...	...	14/1/21
Hamill, J.	...	20020	...	Corporal	...	1st	...	...	21/10/19
Hamilton, P.	...	11291	...	Corporal	...	1st	...	...	14/5/19
Hanley, P.	...	12968	...	L./Corporal	...	2nd	...	...	17/6/19
Bar to M.M.	...	—		—		—			20/8/19
Hanlon, P.	...	27757	...	Private	...	1st	...	...	21/10/18
Hardy, F.	...	5698	...	Private	...	2nd	...	...	23/7/19
Harrington, M.	...	19062	...	Private	...	1st	...	...	17/4/17
Hart, W. T.	...	3/12005	...	Private	...	9th	...	...	21/9/16
Hayes, R.	...	5780	...	Private	...	1st	...	...	11/11/16
Heeney, N.	...	23928	...	Pte. (A./L./Cpl.)	...	1st	...	...	14/5/19
Herbert, J. J.	...	19005	...	Sergeant	...	1st	...	...	11/2/19
Hession, A.	...	9344	...	Private	...	2nd	...	...	13/3/18
Hickey, J.	...	5573	...	Private	...	1st	...	...	11/2/19
Hoey, J.	...	19593	...	L./Corporal	...	8th	...	...	10/8/16
Hogan, J.	...	16811	...	Sergeant	...	6th	...	...	9/12/16
Holmes, W.	...	5433	...	Sergeant	...	2nd	...	...	11/11/16
Hooper, E.	...	24294	...	Private	...	1st	...	...	11/2/19
Hooten, J.	...	28493	...	Corporal	...	2nd	...	...	17/6/19
Horton, H.	...	14870	...	Corporal	...	8th	...	...	26/3/17
Bar to M.M.	...	—		—		—			21/8/17
Hosie, W.	...	28161	...	Private	...	1st	...	...	17/6/19
Hubbard, J. T.	...	40247	...	Private	...	1st	...	...	11/2/19
Humphreys, H. W.	...	14372	...	Private	...	6th	...	...	17/6/19
Humphries, G. E.	...	41390	...	Private	...	1st	...	...	11/2/19
Inglesant, J.	...	29637	...	Private	...	6th	...	...	17/6/19
Irwin, T. F. A.	...	29522	...	Pte. (A./L./Cpl.)	...	2nd	...	...	23/7/19
Jeeves, V.	...	14391	...	L./Corporal	...	7th	...	...	6/1/17
Joceylinn, P.	...	11156	...	Private	...	2nd	...	...	19/11/17
Jones, W.	...	22916	...	Corporal	...	8th	...	...	26/3/17
Joyce, H.	...	27192	...	Private	...	10th	...	...	19/2/17
Kearns, P.	...	23678	...	Private	...	2nd	...	...	13/3/18
Keegan, T.	...	10880	...	A./Sergt.	...	8th	...	...	16/11/16
Keeping, B. R.	...	10946	...	Sergeant	...	2nd	...	...	13/3/18
Keeping, F. A.	...	9497	...	A./Corporal	...	8th	...	...	6/1/17
Kelly, J.	...	15533	...	Private	...	9th	...	...	14/9/16
Kelly, J. P.	...	14653	...	Sergeant	...	9th	...	...	16/11/16
Kelly, P.	...	8790	...	Private	...	1st	...	...	14/1/18
Kenna, E.	...	21076	...	Private	...	2nd	...	...	17/6/19
Kenna, G.	...	23219	...	Private	...	2nd	...	...	23/7/19
Kennedy, D.	...	18324	...	A./Sergeant	...	2nd	...	...	18/7/17
Kenny, W.	...	7359	...	Sergeant	...	2nd	...	...	18/7/17
Killeen, J.	...	19879	...	Private	...	2nd	...	...	23/7/19
Killeen, P.	...	11570	...	Private	...	2nd	...	...	11/11/16
King, C. J.	...	20079	...	L./Corporal	...	8th	...	...	11/5/17
King, H.	...	30183	...	Corporal	...	1st	...	...	17/6/19
Kinsella, J.	...	9891	...	Cpl. (Sergt.)	...	9th	...	...	19/11/17
Leonard, J.	...	25241	...	Private	...	10th	...	...	22/11/19

NAME.	REGTL. NO.	RANK.	BATTALION.	"LONDON GAZETTE."
McCann, P.	7909	Private	8th	19/11/17
McCarten, J.	21563	Pte. (L./Cpl.)	1st	11/2/19
McCarthy, J.	43067	Corporal	1st	9/7/17
McCarthy, J.	16762	Sergeant	1st	14/1/18
McCauley, W.	18856	Sergeant	8th	10/8/16
McCormack, J. J.	27020	Sergeant	10th	19/2/17
McCormick, C.	20910	Private	1st	11/2/19
McCudden, F.	14863	L./Corporal	9th	17/9/17
McCue, E.	18645	L./Corporal	9th	14/9/16
McCullagh, G.	26447	Corporal	10th	13/3/18
McCullagh, M.	9314	Sergeant	1st	11/2/19
McDonnell, J.	19164	Corporal	1st	17/6/19
Bar to M.M.	—	—	—	20/8/19
McFarlane, J.	10814	Corporal	Late 1st	21/9/16
McGahey, G.	7069	Sergeant	2nd	14/12/16
McGann, T.	22117	Private	1st	11/11/16
McGeeney, M.	18624	Pte. (L./Cpl.)	1st	25/4/18
McGlynn, J.	20009	Private	2nd	11/11/16
McGrane, M.	8456	Private	2nd	13/3/18
McGrogan, P.	22141	Private	2nd	13/3/18
McGuirk, D.	21338	Pte. (L./Cpl.)	2nd	13/3/18
McInespie, P.	18241	Private	2nd	19/11/17
Mackew, A.	14090	Sergeant	2nd	14/5/19
McKinley, J. E.	13943	Sergeant	8th	10/8/16
McLoughlin, M.	25245	Cpl. (Sergt.)	10th	13/3/18
McLoughlin, P.	25899	Private	8th	11/5/17
McManus, J.	19870	Private	2nd	29/8/18
McManus, J.	40313	Sergeant	1st	11/2/19
McNeely, J.	23998	Sergeant	1st	17/6/19
McNicol, R.	21890	A./Corporal	1st	11/11/16
McPartlin, M.	21379	Sergeant	2nd	6/8/18
Bar to M.M.	—	—	—	21/10/18
McQueen, J.	21378	Private	1st	17/4/17
Madden, A.	25710	Private	9th	17/9/17
Bar to M.M.	—	—	—	19/10/17
Maguire, P.	8751	Private	Late 1st	11/11/16
Maher, M.	17191	Private	1st	14/1/18
Maher, M.	8/23023	Corporal	8th	14/9/16
Maher, P.	17298	L./Corporal	2nd	17/6/19
Mallon, L.	22461	Pte. (L./Cpl.)	2nd	19/11/17
Bar to M.M.	—	—	—	14/5/19
Malone, P.	22354	Private	1st	28/7/17
Mangan, P. J.	8/16540	Sergeant	8th	14/9/16
Manning, R.	9020	Private	1st	11/2/19
Marchant, W. G.	7/28515	Sergeant	7th	24/1/19
Bar to M.M.	—	—	—	20/8/19
Marks, T.	14116	L./Corporal	2nd	14/5/19
Marshall, A. W.	43546	Corporal	2nd	17/6/19
Martin, J.	15543	Private	8th	9/12/16
May, J. P.	18196	Private	7th	6/1/17
Mealey, L. P.	21136	Private	2nd	23/7/19
Meehan, J.	9138	Private	2nd	19/11/17
Merrins, J.	6568	Sergeant	2nd	4/2/18
Bar to M.M.	—	—	—	2/4/18
Miller, J.	23815	Private	2nd	19/11/17
Minogue, J. P.	19274	Private	2nd	23/7/19
Mitchell, A. J.	24717	Private	10th (attd. 54th L.R.O. Coy. R.E.)	19/11/17
Money, J. W.	43216	Pte. (A./Sgt.)	2nd (attd. 48th T.M. Bty.)	19/11/17

NAME.		REGTL. NO.		RANK.		BATTALION.			"LONDON GAZETTE."
Mooney, M. ...	...	6978	...	Sergeant	...	9th	...	...	19/11/17
Moore, P. ...	...	8272	...	Private	...	1st	...	...	11/11/16
Moran, J. ...	...	13008	...	Corporal	...	1st (attd. 48th Inf. Bde. Sig. Sec. R.E.)			29/8/18
Moran, T. ...	...	26925	...	Private	...	10th	...	...	26/5/17
Morrisey, T. ...	...	23002	...	Corporal	...	1st	...	...	28/9/17
Bar to M.M.	...	—		—		—			14/1/18
Mortimer, F. J.	...	30108	...	Private	...	2nd	...	...	23/7/19
Mulhall, P. ...	...	27634	...	Pte. (L./Cpl.) ...		1st	...	...	30/1/20
Mulhall, T. ...	...	6/12082	...	L./Corporal ...		Late 6th	...		26/4/17
Mullen, A. ...	...	15074	...	Sergeant	...	2nd	...	...	20/8/19
Mullen, M. ...	...	40055	...	Private	...	2nd	...	...	13/3/18
Mulligan, E. ...	...	7163	...	Private	...	1st	...	...	14/1/18
Mulligan, F., D.C.M.		5834	...	Private	...	1st	...	...	14/5/19
Bar to M.M.	...	—		—		—			23/7/19
Mulrenan, T.	...	25970	...	Private	...	10th	...	...	26/4/17
Murphy, J. ...	...	21612	...	Cpl. (A./Sgt.) ...		8th	...	...	19/11/17
Murphy, J. ...	...	26136	...	Private	...	1st	...	...	11/2/19
Murphy, J. ...	...	15061	...	Private	...	1st	...	...	11/2/19
Murray, F., D.C.M. ...		11371	...	Private	...	2nd	...	...	14/5/19
Murray, J. ...	...	27979	...	Pte. (A./L./Cpl.)		1st	...	...	14/5/19
Murray, J. ...	...	17074	...	Private	...	1st	...	...	17/6/19
Murray, J. ...	...	26629	...	Private	...	8th	...	...	19/11/17
Murray, M. ...	...	10300	...	Private	...	2nd	...	...	29/8/18
Bar to M.M.	...	—		—		—			17/6/19
Navin, T. ...	...	24605	...	Private	...	1st	...	...	9/7/17
Neville, J. ...	...	25857	...	Private	...	1st	...	...	9/7/17
Newell, A. ...	...	24219	...	Private	...	2nd	...	...	17/6/19
Newman, W. ...	...	13693	...	Corporal	...	9th	...	...	14/9/16
Newton, F. P. ...	...	9211	...	C.S.M.	...	1st	...	...	11/2/19
Bar to M.M.	...	—		—		—			17/6/19
Norfolk, W. V. ...	...	28774	...	L./Corporal ...		2nd	...	...	14/5/19
Norman, R. ...	...	28018	...	Pte. (A./L./Cpl.)		1st	...	...	11/2/19
Normoyle, M. ...	...	17239	...	Corporal	...	9th	...	...	19/11/17
O'Brien, J. ...	...	10055	...	L./Corporal ...		1st	...	...	3/6/16
O'Brien, J. ...	...	13886	...	Sergeant	...	9th	...	...	16/11/16
O'Brien, J., D.C.M. ...		9581	...	Corporal	...	1st	...	...	11/2/19
Bar to M.M.	...	—		—		—			17/6/19
O'Brien, P. ...	...	10924	...	L./Corporal ...		1st	...	...	3/6/16
O'Brien, R. ...	...	23727	...	Private	...	1st	...	...	20/10/19
O'Byrne, L. ...	...	6189	...	A./Sergeant ...		1st	...	...	11/11/16
O'Connell, M. ...	...	26174	...	L./Corporal ...		10th	...	...	19/2/17
O'Connor, H. ...	...	12015	...	Cpl. (A./L./Sgt.)		2nd	...	...	19/11/17
O'Donnell, C. ...	...	7124	...	L./Corporal ...		1st	...	...	11/11/16
O'Haire, P. ...	...	8598	...	Private	...	1st	...	...	25/4/18
O'Neill, T. ...	...	21339	...	L./Corporal ...		9th	...	...	10/8/16
O'Toole, D. ...	...	25501	...	Private	...	10th	...	...	19/2/17
Ockenden, J. ...	...	10605	...	Sergeant	...	1st	...	...	28/9/17
Owen, A. ...	...	28776	...	Private	...	2nd	...	...	23/7/19
Parsons, G. ...	...	8056	...	Private	...	5th	...	...	24/1/17
Patterson, W. P.	...	30501	...	Private	...	1st	...	...	11/2/19
Peters, A. E. ...	...	6/14311	...	Corporal	...	6th	...	...	2/11/17
Petty, H. E. ...	...	28221	...	Private	...	8th	...	...	19/11/17
Pim, J. ...	...	10410	...	Pte. (A./L.Cpl.)		1st	...	...	11/2/19
Powell, T. ...	...	18277	...	Private	...	2nd	...	...	17/6/19
Preece, G. R. ...	...	25437	...	Sergeant	...	10th	...	...	24/1/17
Prestage, R. ...	...	10347	...	Sergeant	...	1st	...	...	18/10/17
Price, H. ...	...	26059	...	Private	...	9th	...	...	19/11/17
Price, J. ...	...	10616	...	Private	...	2nd	...	...	19/11/17
Priest, T. ...	...	25462	...	Sergeant	...	10th	...	...	19/2/17

NAME.		REGTL. NO.		RANK.		BATTALION.			"LONDON GAZETTE."
Pudney, E. J.	...	31272	...	Private	...	6th	...	...	20/10/19
Quinn, J.	...	8/15102	...	Private	...	8th	...	...	14/9/16
Quinn, J.	...	12894	...	Private	...	9th	...	...	19/11/17
Redmond, W.	...	10722	...	Private	...	10th	...	...	26/4/17
Regan, C. C.	...	21097	...	Private	...	1st	...	...	11/2/19
Regan, J.	...	24016	...	Private	...	2nd	...	...	17/6/19
Reid, G.	...	21584	...	Private	...	2nd	...	...	17/6/19
Reilly, P.	...	9042	...	Cpl. (L./Sgt.)	...	9th	...	...	19/11/17
Reilly, T.	...	27045	...	Private	...	1st	...	...	17/6/19
Rex, P. C.	...	9025	...	Private	...	1st	...	...	3/6/16
Roache, A.	...	5068	...	Private	...	2nd	...	...	11/11/16
Roberts, J. F.	...	16878	...	Corporal	...	2nd	...	...	18/7/17
Bar to M.M.	...	—		—		—			13/3/18
Robinson, J.	...	18870	...	Private	...	9th	...	...	19/11/17
Roe, P.	...	11416	...	Private	...	2nd	...	...	3/6/16
Roebuck, J.	...	28547	...	Pte. (L./Cpl.)	...	2nd	...	...	17/6/19
Rogers, J.	...	40312	...	Corporal	...	1st	...	...	14/1/18
Ryan, P.	...	43135	...	Private	...	6th	...	...	20/8/19
Ryder, J.	...	25794	...	Private	...	1st	...	...	11/2/19
Sandar, W.	...	24040	...	Private	...	2nd	...	...	13/3/18
Sargeant, T.	...	18773	...	Private	...	1st	...	...	6/8/18
Saunders, J.	...	7842	...	Private	...	1st	...	...	14/1/18
Scallon, W.	...	19191	...	Private	...	9th	...	...	16/11/16
Scott, J.	...	14095	...	Private	...	8th	...	...	17/9/17
Scully, J.	...	12209	...	Private	...	1st	...	...	14/5/19
Bar to M.M.	...	—		—		—			23/7/19
Seddon, G.	...	22177	...	Private	...	1st	...	...	14/1/18
Shinkwin, T.	...	8775	...	Corporal	...	2nd	...	...	18/7/17
Short, E. J.	...	25975	...	Private	...	10th	...	...	26/5/17
Singleton, J.	...	26941	...	Pte. (L./Cpl.)	...	2nd	...	...	23/7/19
Skegg, C.	...	14306	...	L./Sergeant	...	6th	...	...	17/6/19
Skinner, F. C.	...	28560	...	L./Corporal	...	2nd	...	...	17/6/19
Smith, J.	...	9834	...	Private	...	2nd	...	...	19/11/17
Smith, M.	...	8145	...	Private	...	2nd	...	...	13/3/18
Smith, R.	...	14548	...	Private	...	8th	...	...	10/8/16
Sneddon, R.	...	7165	...	Private	...	2nd	...	...	17/6/19
Spratt, E. A.	...	26721	...	Private	...	10th	...	...	19/2/17
Bar to M.M.	...	—		—		—			13/3/18
Stowe, T.	...	27686	...	Private	...	2nd	...	...	17/6/19
Street, L.	...	14384	...	L./Corporal	...	8th	...	...	10/8/16
Tait, T.	...	14613	...	Sergeant	...	8th	...	...	10/8/16
Taylor, G.	...	43203	...	Private	...	1st	...	...	14/5/19
Territt, J.	...	10026	...	Private	...	1st	...	...	18/10/17
Bar to M.M.	...	—		—		—			14/1/18
Thomas, A.	...	11542	...	Sergeant	...	1st	...	...	11/2/19
Thurlow, W.	...	21731	...	Sergeant	...	6th	...	...	17/6/19
Tierney, J.	...	21170	...	Private	...	1st	...	...	11/11/16
Townsend, F.	...	25184	...	Private	...	1st	...	...	18/10/17
Tracey, T.	...	18929	...	Private	...	1st	...	...	21/9/16
Tracy, J.	...	9622	...	Private	...	2nd	...	...	11/11/16
Turton, W.	...	3/19688	...	Corporal	...	7th	...	...	25/4/18
Valentine, P.	...	11163	...	Sergeant	...	1st	...	...	11/11/16
Wade, J.	...	9240	...	Sergeant	...	1st	...	...	17/6/19
Walker, H.	...	40726	...	Private	...	1st	...	...	11/2/19
Bar to M.M.	...	—		—		—			17/6/19
Wallace, C. N.	...	6642	...	Sergeant	...	2nd	...	...	6/1/17
Walls, J.	...	12137	...	A./Sergeant	...	1st	...	...	11/2/19
Bar to M.M.	...	—		—		—			23/7/19
Walsh, J.	...	13269	...	Private	...	2nd	...	...	12/12/17
Walsh, M.	...	6745	...	Sergeant	...	2nd	...	...	11/11/16
Watson, J.	...	41004	...	Private	...	1st	...	...	18/10/17

NAME.	REGTL. NO.	RANK.	BATTALION.	"LONDON GAZETTE."
Webb, J.	9072	Private	2nd	18/7/17
Wharf, E.	40388	Private	1st	11/2/19
Whelan, P.	10776	A./Corporal	1st	11/11/16
White, H.	9980	Pte. (L./Cpl.)	2nd	13/3/18
Whittaker, L.	9709	Sergeant	8th	11/5/17
Wilkinson, T.	19115	Private	1st	17/6/19
Wilson, J. H.	15901	Sergeant	9th	14/9/16
Wood, A.	10652	C.Q.M.S.	2nd	3/6/19
Worrell, W.	13119	Private	8th	21/9/16

MERITORIOUS SERVICE MEDAL.

NAME.	REGTL. NO.	RANK.	BATTALION.	"LONDON GAZETTE."
Berry, W. P.	8898	Sergeant	2nd	17/6/18
Boon, W.	10464	C.Q.M.S.	2nd	30/1/20
Burke, J.	10809	C.Q.M.S.	2nd	3/6/19
Cahill, J.	9829	Sgt. (A./C.Q.M.S.)	2nd	3/6/19
Cameron, J.	15955	Sergeant	1st	3/6/19
Clements, W. H.	7/14401	Private	7th	26/4/17
Connolly, R. W.	15825	C.S.M.	10th	17/6/18
Cops, F. O.	12123	R.Q.M.S.	1st	17/6/18
Cornish, J.	12170	Sergeant	6th	3/6/19
Coyne, J.	7/13760	C.S.M.	7th	13/2/17
Daly, J.	4956	R.S.M.	—	22/2/19
Darling, S. G.	7/15273	Sergeant	7th	26/4/17
Fitzpatrick, F. H.	29404	Q.M.S. (A./S.M.)	—	12/12/19
Frost, G. F.	6/12325	R.S.M.	6th	18/6/17
Galloway, E. H.	25356	Sgt. (A./C.S.M.)	—	18/1/19
Geraghty, H.	8658	Sergeant	1st	17/6/18
McKeown, D.	15117	R.Q.M.S.	8/9th	17/6/18
Mangan, W.	6/13553	C.Q.M.S.	6th	17/12/17
Marrett, E. A. R.	43537	Pte. (A./Q.M.S.)	2nd	3/6/19
Money, J. W.	43216	Sergeant	2nd (attd. 48th T.M. Bty.)	17/6/18
O'Connor, J. J.	8859	R.Q.M.S.	2nd	17/6/18
Purcell, P.	15120	Q.M.S.	Depot	12/12/19
Shoetensack, W. G.	28557	Private	7th	3/6/19

MENTIONED IN DESPATCHES, 1914–20.

NAME.	RANK.	BATTALION.	THEATRE OF WAR.	"LONDON GAZETTE."
Atherley, C. E.	Temp. Lieut. (Service Bn. Yorkshire L.I.)	Attd. 1st	Dardanelles	5/11/15, p. 10999
Allan, J. H.	Capt. (Temp. Major)	1st attd. 1/7th Manch.	Egypt	6/7/17, p. 6769
Addis, G. T.	Capt. (Spec. Res.)	4th (attd. 2nd)	France	21/12/17, p. 13373
			France	24/5/18, p. 6101
Adcock, H.	Pte. (Regtl. No. 27802)	6th	France	9/7/19, p. 8710
Bousted, L. C.	Lieut.	1st	Dardanelles	5/11/15, p. 10999
Brennan, T.	L./Cpl. (Regtl. No. 5610)	2nd	France	1/1/16, p. 56
Bryan, E.	Cpl. (Regtl. No. 13197)	Service	Dardanelles	28/1/16, p. 1204
Burrowes, A. E.	Sergt. (Regtl. No. 14150)	Service	Dardanelles	28/1/16, p. 1204
Bellingham, E. H. C. P., D.S.O.	Temp. Lieut.-Col. (Lieut. R. of O., Temp. Brig.-Gen.)	8th	France	4/1/17, p. 240
			France	11/12/17
Byrd, B.	Sergt. (A./C.S.M.) (Regtl. No. 16517)	8th	France	4/1/17, p. 240
Burke, F. W. R.	L./Sergt. (Regtl. No. 25692)	10th	—	25/1/17, p. 946
Brown, F. E.	Lieut.	3rd (attd. R.F.C.)	France	15/5/17, p. 4754
Burke, J., M.C., D.C.M.	Qr.Mr. and Hon. Major	2nd	France	17/2/15, p. 1668
			France	21/12/17, p. 13373
			France	9/7/19, p. 8710
Burke, J.	Sergt. (Regtl. No. 6/14885)	6th	Egypt	14/6/18, p. 7054
Byrne, R. C., M.C.	Qr.Mr. and Hon. Capt.	6th	Dardanelles	28/1/16, p. 1204
			Egypt	14/6/18, p. 7054
Beveridge, E. W.	Lieut. (attd. 34th D.S. Coy. R.E.)	4th	France	9/7/19, p. 8710
Byrne, L. C., D.S.O., M.C.	Lieut. (A./Major)	2nd	France	9/7/19, p. 8710
Bailey, H. T.	Cpl. (A./L./Sergt.) (Regtl. No. 15372)	2nd	France	9/7/19, p. 8710
Brown, C. H.	Sergt. (Regtl. No. 43472)	1st	France	9/7/19, p. 8710
Boon, W., M.M.	C.Q.M.S. (Regtl. No. 10464)	2nd (Wrexham)	—	30/1/20, p. 1227
Campbell, F. C. G.	Lieut. (40th Pathans)	Attd. 2nd	France	17/2/15, p. 1668
Cooke, —	Cpl. (Regtl. No. 8672)	2nd	France	17/2/15, p. 1668
Curley, W. P.	Cpl. (Regtl. No. 9508)	2nd	France	22/6/15, p. 6003
Crozier, H. C.	Capt.	1st	Dardanelles	5/8/15, p. 7666
Cooney, C.	Sergt. (Regtl. No. 10256)	1st	Dardanelles	5/8/15, p. 7666
Cummins, W.	Sergt. (Regtl. No. 6603)	1st	Dardanelles	5/8/15, p. 7666
			Dardanelles	5/5/16, p. 4519
Carew, R. J. H.	Capt. (Temp. Major, 83rd Bde. Machine Gun Officer)	2nd	France	1/1/16, p. 56
			Salonika	11/6/18, p. 6918
			Salonika	5/6/19

NAME.	RANK.	BATTALION.	THEATRE OF WAR.	"LONDON GAZETTE."
Campbell, J.	Temp. Sergt.-Major (Regtl. No. 13507), also Qr.Mr. and Hon. Lieut.	8th	Dardanelles	28/1/16 p. 1204
			France	25/5/17, p. 5166
			France	21/12/17
Carruthers, C. G.	Capt.	1st	Dardanelles	5/11/15, p. 10999
			Dardanelles	5/5/16, p. 4519
			Dardanelles	13/7/16, p. 6956
Clarke, N. P.	Major	2nd	France	17/2/15, p. 1668
			Egypt	21/6/16, p. 6182
Cory, G. N.	Major and Bt. Col. (Temp. Major-Gen.)	R.D.F., Staff	France	9/12/14, p. 10535
			France	17/2/15, p. 1648
			France	1/1/16, p. 3
			Salonika	25/9/16, p. 9342
			Salonika	6/12/16
			Salonika	21/7/17, p. 7448
			Salonika	11/6/18, p. 6918
			Salonika	31/1/19, p. 1474
			Mesopotamia	9/9/21
Cox, P. G. A.	Bt. Major (Temp. Lieut.-Col.), Ret. Pay, R. of O.	6th	Dardanelles	28/1/16, p. 1204
			Salonika	13/7/16, p. 6942
			Salonika	6/12/16
Corcoran, J.	Pte. (Regtl. No. 15632)	8th	France	4/1/17, p. 240
Corballis, E. R. L.	Capt.	2nd (attd. R.F.C. Staff)	France	17/2/15
			France	15/5/17, p. 4754
			France	11/12/17, p. 12911
Culbert, W.	Sergt. (A./Q.M.S.) (Regtl. No. 10365)	1st	France	25/5/17, p. 5166
Cahill, W.	Sergt. (A./Q.M.S.) (Regtl. No. 18859)	8th	France	25/5/17, p. 5166
Cullen, A. A.	Temp. 2nd-Lieut.	6th	Salonika	21/7/17, p. 7453
Curtis, H. A.	Cpl. (A./Sergt.) (Regtl. No. 7/14107)	7th	Salonika	21/7/17, p. 7454
Curran, P.	R.Q.M.S. (Regtl. No. 1/9593)	7th	Salonika	28/11/17, p. 12485
Conarchy, W. C.	Temp. 2nd-Lieut.	8th	France	21/12/17, p. 13373
Cowley, G. E.	Capt. (Temp. Major)	Attd. 8th	France	21/12/17, p. 13373
Cameron, J.	Sergt. (Regtl. No. 15955)	1st	France	21/12/17, p. 13373
Cops, F. O.	R.Q.M.S. (Regtl. No. 12123)	1st	France	21/12/17, p. 13373
			France	9/7/19, p. 8710
Considine, T. J.	Capt.	5th (attd. 1st)	France	28/12/18, p. 15162
Coombs, J. H.	Sergt.-Dmr. (Regtl. No. 28826)	2nd	France	9/7/19, p. 8710
Dickie, J. MacN.	Lieut.	2nd	France	22/6/15, p. 6003

Name	Rank	Battalion	Theatre	Gazette
Devoy, J.	Sergt. (Regtl. No. 10335)	7th	Dardanelles	28/1/16, p. 1204
Downing, G.	Lieut.-Col. (R. of O.)	7th	Dardanelles	28/1/16, p. 1204
Dyke, C. P.	Pte. (Regtl. No. 9761)	—	Dardanelles	5/5/16, p. 4519
Daly, U. de B.	Capt.	4th	Home "A" List	25/1/17, p. 944
Doran, W. A.	Temp. Capt.	9th	France	25/5/17, p. 5166
Dee, G. S.	Temp. 2nd-Lieut.	7th	Salonika	28/11/17, p. 12485
Douglas, R. G.	Lieut. (Temp. Capt.)	2nd	Egypt	16/1/18, p. 934
	Staff (Temp. Major)		Egypt	5/6/19, p. 7172
Delaney, A. S.	Lieut. (Temp. Capt.)	5th (attd. 2nd)	France	24/5/18, p. 6101
Eales, R. W.	Pte. (Regtl. No. 7/15351)	7th	Salonika	6/12/16, p. 11936
Esmonde, L. G.	Lieut.-Col.	—	(List "A," Ireland)	25/1/17, p. 944
Esmonde, James, M.C.	Lieut. (A./Capt.)	Attd. 6th	Salonika	28/11/17, p. 12485
			France	9/7/19, p. 8710
Ekins, E. W.	Pte. (Regtl. No. 28873)	2nd	France	9/7/19, p. 8710
Frankland, T. H. C.	Capt.	2nd	France	17/2/15, p. 1668
Ferguson, S.	Sergt. (Regtl. No. 6128)	1st	Dardanelles	5/8/15, p. 7666
Fletcher, D. A.	Lieut. (Temp. Capt.) 2nd Cameron Highrs. Attd. R.D.F.		Salonika	6/12/16, p. 11939
Finnigan, J.	Pte. (Regtl. No. 43128)	—	Home "A" List	25/1/17, p. 946
Forbes, A.	Sergt. (Regtl. No. 7/14774)	7th	Salonika	28/11/17, p. 12485
Fitzgerald, P.	C.Q.M.S. (A./C.S.M.) (Regtl. No. 9952)	1st	France	9/7/19, p. 8710
Graham, A.	C.S.M. (Regtl. No. 6203)	2nd	France	22/6/15, p. 6003
Guest, A.	R.S.M. (Regtl. No. 14153)	7th	Dardanelles	28/1/16, p. 1204
	Temp. 2nd-Lieut.	—	Salonika	28/11/17, p. 12485
Gibbs, G. H.	2nd-Lieut. (A./Capt.)	4th	France	25/5/17, p. 5166
Grove, J. R. W.	Capt. (Temp. Major)	1st (attd. R.F.C.)	North Russia	16/1/19, p. 823
Glegg, J. D., M.C.	Capt.	2nd, Staff	France	5/7/19, p. 8492
Gibbs, J. L. A.	Lieut. (A./Capt.)	4th (attd. 5th Field Surv. Bn., R.E.)	France	9/7/19, p. 8710
Gilder, A. W.	Sergt. (Regtl. No. 40518)	2nd (attd. H.Q. 48th Inf. Bde.)	France	9/7/19, p. 8710
Hall, R. S.	C.S.M. (Regtl. No. 5039)	2nd	France	19/10/14, p. 8356
Hatt, F. W.	Sergt.-Major (Regtl. No. 5531)	2nd	France	17/2/15, p. 1668
	R.S.M.	2nd	France	24/5/18, p. 6101
Henderson, E.	C.S.M. (Regtl. No. 10819)	2nd	France	17/2/15, p. 1668
Hayes, R.	Pte. (Regtl. No. 5780)	1st	Dardanelles	5/11/15, p. 10999
Hayden, P.	Pte. (Regtl. No. 9470)	2nd	France	1/1/16, p. 56
Harrison, R.S.M.	Major 51st Sikhs	Attd. 7th	Dardanelles	28/1/16, p. 1204
Hoey, C. B. R.	Capt. (Temp. Major)	1st (attd. 7th)	Dardanelles	28/1/16, p. 1204

(See also page 198.)

NAME.	RANK.	BATTALION.	THEATRE OF WAR.	"LONDON GAZETTE."
Haig, T.	C.S.M. (Regtl. No. 14972)	7th	Dardanelles	28/1/16, p. 1204
Higginson, H. W., C.B., D.S.O.	Major and Bt. Col. (Temp. Major-Gen.)	R.D.F. and Staff*	France	1/1/16, p. 6
			France	15/6/16, p. 5951
			France	4/1/17
			France	15/5/17, p. 4748
			France	11/12/17, p. 12914
			France	20/5/18, p. 5947
			France	20/12/18, p. 14928
			France	5/7/19, p. 8493
Harvey, C. D.	Temp. Lieut.	7th	Salonika	6/12/16, p. 11939
Haskard, J. McD., C.M.G., D.S.O.	Major (Temp. Brig.-Gen.)	Staff	France	4/1/17
			France	15/5/17, p. 4748
			France	11/12/17, p. 12914
			France	20/5/18, p. 5947
			France	20/12/18, p. 14928
Heffernan, J. G. P.,	Temp. Capt.	8/9th (attd. 1st)	France	4/1/17, p. 240
	Temp. Major	—	France	24/5/18, p. 6101
			France	15/7/19, p. 8710
Hunt, J. P., C.M.G., D.S.O., D.C.M.	Temp. Major (A./Lieut.-Col.)	Attd. 8th	France	4/1/17, p. 240
			France	21/12/17, p. 13373
	Gen. List	—	France	28/12/18, p. 15162
	Gen. List, Comdg. 1st Bn. R. Irish Rifles	—	France	10/7/19, p. 17
Holmes, W.	Sergeant (Regtl. No. 5433)	2nd	France	4/1/17, p. 240
Henderson, J. S.	C.S.M. (Regtl. No. 14779)	10th	Home "A" List	25/1/17, p. 946
	A./R.S.M.	—	France	25/5/17, p. 5166
Holloway, L.	Hon. Capt. and Qr.Mr.	7th	Salonika	21/7/17, p. 7454
Harney, J.	Pte. (Regtl. No. 27952)	10th	France	21/12/17, p. 13373
Hamlet, G. T.	Temp. Capt.	10th	France	24/5/18, p. 6101
Harrison, E. N.	C.Q.M.S. (Regtl. No. 7/28484)	7th	Egypt	14/6/18, p. 7054
Halligan, J. T.	Capt. (2nd-Lieut. late Leinster Regt. ; Adjt. and Qr.Mr. Cadet School)	—	France	28/12/18, p. 15162
Harcourt, H. G., D.S.O., M.C.	Lieut. (A./Major)	Attd. 51st Bn. M.G. Corps	France	28/12/18, p. 15162
			Archangel	11/6/20, p. 6456
Hawes, C. E.	Temp. Lieut. (Temp. Major)	Staff	France	15/5/17
			France	5/7/19, p. 8493
Ingoldby, R. H.	Temp. 2nd-Lieut.	2nd	France	1/1/16, p. 56
Johnson, R. D.	Major	3rd (attd. 2nd)	France	22/6/15, p. 6003

Name	Rank				Battalion		Theatre		Date and Gazette
Jennings, T.	Pte. (Regtl. No. 10641) ...	...	...	—			Dardanelles	...	5/5/16, p. 4519
Jeffreys, R. G. B. ...	Major (A./Lieut.-Col.) ...	...	...	1st (attd. 2nd)	...		France ...	...	4/1/17, p. 240
							France ...	...	25/5/17, p. 5166
							France ...	...	21/12/17, p. 13373
Jeffries, W. F., D.S.O. ...	Capt. (A./Major) ...	...	...	3rd	...		France ...	...	4/1/17
							France ...	...	28/12/18, p. 15162
Jones, H.	Pte. (Regtl. No. 10944) ...	...	...	6th	...		France ...	...	9/7/19, p. 8710
Keegan, M.	2nd-Lieut.	...	...	R.D.F. (attd. R.F.C.)			France ...	...	4/1/17, p. 204
Kendrick, E. H., D.S.O.	Capt., Bt. Major (Temp. Lieut.-Col.)		...	Attd. 11th Bn. Suff.			France ...	...	25/5/17, p. 5166
				Regt. and 34th			France ...	...	21/12/17, p. 13373
				Bn. M.G. Corps			France ...	...	28/12/18, p. 15162
							France ...	...	9/7/19, p. 8710
Kee, W., M.C. ...	C.S.M. (Regtl. No. 14132) ...	...	...	7th (attd. 1st)	...		Dardanelles	...	28/1/16, p. 1204
	Temp. Lieut.	...	...	7th (attd. 1st)	...		France ...	...	24/5/18, p. 6101
Lonsdale, M. P. E. ...	Major (R. of O.) ...	...	...	7th	...		Dardanelles	...	28/1/16, p. 1204
Lynch, C.	C.S.M. (Regtl. No. 25573) ...	...	...	10th ...	...		Dardanelles	...	28/1/16, p. 1204
							Home ...	...	25/1/17, p. 946
Leahy, T. J., D.S.O., M.C.	Capt., Bt. Major ...	...	...	2nd	...		France ...	...	19/10/14, p. 8356
							France ...	...	9/12/14, p. 10546
							France ...	...	15/6/15
							France ...	...	1/1/16, p. 56
							France ...	...	4/1/17, p. 198
							France ...	...	15/5/17, p. 4749
							France ...	...	11/12/17, p. 12916
							France ...	...	20/5/18, p. 5948
							France ...	...	20/12/18, p. 14929
Losty, P. P. ...	C.Q.M.S. (Regtl. No. 2/10034)	...	...	6th	...		Salonika ...	...	28/11/17, p. 12485
Lawrence, L. A. ...	Lieut. (A./Capt.) ...	...	...	4th (attd. 9th)	...		France ...	...	21/12/17, p. 13373
Lunn, J. S. ...	Capt.	...	...	4th	...		Egypt ...	...	14/6/18, p. 7054
							Egypt ...	...	12/1/20, p. 504
Langley, B. ...	Sergt. (Regtl. No. 10645) ...	...	...	1st	...		France ...	...	28/12/18, p. 15162
Loveband, A., C.M.G. ...	Lieut.-Col.	...	...	2nd	...		France ...	...	17/2/15, p. 1668
Loveband, G. Y. ...	Temp. Capt.	...	...	6th	...		Egypt ...	...	22/1/19, p. 1156
Lloyd-Blood, L. I. N., M.C.	Lieut.	...	...	5th (attd. 2nd)	...		France ...	...	9/7/19, p. 8710
Loydall, J. ...	Sergeant (No. 16646)	...	...	2nd	...		France ...	...	25/5/17
Massy-Westropp, R. F. H.	Lieut.	...	...	2nd	...		France ...	...	4/12/14, p. 10296
Murphy, P. ...	Cpl. (Regtl. No. 7755) ...	...	...	2nd	...		France ...	...	22/6/15, p. 6003
Molesworth, E. A. ...	Major	...	...	1st	...		Dardanelles	...	5/8/15, p. 7666
Maclear, B.	Capt.	...	...	2nd	...		France ...	...	1/1/16, p. 56

* Was serving with 2nd Bn. when the first two mentions were awarded.

NAME.	RANK.	BATTALION.	THEATRE OF WAR.	"LONDON GAZETTE."
Mulhall, P.	Pte. (Regtl. No. 10714)	1st	Dardanelles	5/5/16, p. 4519
Murphy, W. H.	2nd-Lieut.	4th	France	4/1/17, p. 240
Murphy, W. J.	Temp. Capt.	9th	France	4/1/17, p. 240
Meldon, J. A.	Lieut.-Col.	4th	Home "A" List	25/1/17, p. 945
MacDaniel, J. R.	2nd-Lieut.	3rd	France	25/5/17, p. 5166
Merry, J.	Qr.Mr. and Hon. Lieut.	9th	France	25/5/17, p. 5166
Macdermott, A. W.	Temp. Capt.	7th	Salonika	21/7/17, p. 7453
Marchant, W. G.	Cpl. (Regtl. No. 7/28515)	7th	Salonika	21/7/17, p. 7454
McAuley, H. T. B.	L./Sergt. (A./Sergt.) (Regtl. No. 40300)	7th (now E. Yorks Regt.)	Salonika	28/11/17, p. 12485
McFeely, C. M., D.S.O., M.C.	Capt.	1st	France	21/12/17, p. 13373
Maguire, R., M.C.	Temp. 2nd-Lieut.	1st	France	21/12/17, p. 13373
Moore, A., D.S.O.	Major	1st	—	—
	Lieut.-Col. New Zealand Infantry	Otago Bn.	Dardanelles	5/11/15, p. 11003
	Major	R.D.F.	France	21/12/17, p. 13373
Murray, A. H.	Temp. 2nd-Lieut. (A./Capt.)	9th	France	25/5/17, p. 5166
			France	21/12/17, p. 13373
Moore, C.	Cpl. (L./Sergt.) (Regtl. No. 8840)	8th	France	24/5/18, p. 6101
Matson, C., M.C.	Lieut. (A./Major)	Secd. 47th Bn. M.G. Corps	France	9/7/19, p. 8710
McCann, C.	Lieut.	3rd (attd. 6th)	France	9/7/19, p. 8710
Maffett, C. W.	Capt.	1st	France	9/7/19, p. 8710
Nolan, E.	Regtl. No. 24910	5th	Home "A" List	25/1/17, p. 946
Neill, C.	Temp. Lieut.	10th (attd. 11th Hants.)	France	24/5/18, p. 6101
O'Hagan, C. E.	Sergt. (Regtl. No. 5753)	2nd Bn.	France	17/2/15, p. 1668
O'Hara, H. D.	Lieut.	1st	Dardanelles	5/8/15, p. 7666
O'Brien, J.	Pte. (Regtl. No. 10055)	1st	Dardanelles	5/11/15, p. 10999
			Dardanelles	5/5/16, p. 4519
O'Brien, P.	L./Cpl. (Regtl. No. 10924)	1st	Dardanelles	5/11/15, p. 10999
			Dardanelles	5/5/16, p. 4519
O'Dowda, J. W., C.B., C.M.G.	Lieut.-Col. and Bt. Col. (Temp. Brig.-Gen.)	1st	Dardanelles	13/7/16, p. 6943
			Mesopotamia	19/10/16, p. 10048
			Mesopotamia	15/8/17, p. 8329
			Mesopotamia	12/3/18, p. 3112
			Mesopotamia	27/8/18, p. 9985
			Mesopotamia	21/2/19, p. 2588
			Mesopotamia	5/6/19, p. 7234

Name	Rank	Battalion	Theatre	Date, page
O'Donnell, M. F.	Sergt. (Regtl. No. 17393)	7th	Salonika	6/12/16, p. 11939
Oulton, W. P., M.C.	Capt.	1st	France	4/1/17, p. 240
O'Neill, F.	2nd-Lieut. (killed)	5th	Home "A" List	25/1/17, p. 945
O'Connor, J. J.	Q.M.S. (Regtl. No. 8859)	2nd	France	22/6/15, p. 6003
	R.Q.M.S.	—	France	25/5/17, p. 5166
O'Malley, V. D., M.C.	Temp. Capt.	10th	France	25/5/17, p. 5166
			France	24/5/18, p. 6101
O'Malley, P.	Pte. (Regtl. No. 6/12871)	6th	Salonika	21/7/17, p. 7454
O'Neill, J.	Temp. Lieut.	8/9th	France	24/5/18, p. 6101
Preston, A. J. D.	Capt.	7th	Dardanelles	28/1/16, p. 1204
Palmer, L. S. N.	Temp. Capt.	7th	Dardanelles	28/1/16, p. 1204
Popham, F. S.	Capt.	5th	Home "A" List	25/1/17, p. 945
Pedlow, W.	2nd-Lieut. (A./Capt.)	2nd	France	21/12/17, p. 13373
Plunkett, J. F., D.S.O., M.C., D.C.M.	Capt. (Temp. Lieut.-Col.)	1st (attd. 13th Bn. R. Innis. Fus.)	France	28/12/18, p. 15162
		Attd. 15th Bn. R. Innis. Fus.	France	9/7/19, p. 8710
Padley, A. C.	Lieut.	9th	Siberia	14/1/20, p. 665
Ray, A. W.	Sergt. (Regtl. No. 6704)	2nd	France	19/10/14, p. 8356
Richards, W. R.	Capt.	6th	Dardanelles	28/1/16, p. 1204
Robinson, H.	C.S.M. (Regtl. No. 14275)	7th	Dardanelles	28/1/16, p. 1204
Rex, P. C.	Pte. (Regtl. No. 9025)	—	Dardanelles	5/5/16, p. 4519
Romer, C. F., C.B., A.D.C.	Lieut.-Col. and Bt.-Col. (Temp. Brig.-Gen.; Major-Gen.)	Staff	France	9/12/14, p. 10535
			France	17/2/15, p. 1650
			France	22/6/15, p. 6010
			France	1/1/16, p. 9
			France	15/6/16, p. 5923
			France	4/1/17, p. 201
			France	15/5/17, p. 4751
			France	11/12/17, p. 12919
Renny, L. F.	Major (Temp. Brig.-Gen.)	Staff	France	17/2/15, p. 1650
			France	22/6/15, p. 5979
			France	1/1/16, p. 9
			France	15/6/16, p. 5923
			France	4/1/17, p. 201
			Home	12/2/18, p. 1934
			France	5/7/19, p. 8498

NAME.	RANK.	BATTALION.	THEATRE OF WAR.	"LONDON GAZETTE."
Robinson, J. P. B., C.M.G., D.S.O.	Major (A./Lieut.-Col.)	Staff	France	4/1/17, p. 201
			France	15/5/17, p. 4751
			France	11/12/17, p. 12919
			France	20/5/18, p. 5950
			France	20/12/18, p. 14933
Rickerby, S.	Cpl. (Regtl. No. 11695)	2nd	France	4/1/17, p. 240
Rawlings, H. W.	L./Cpl. (Regtl. No. 6/14305)	6th	Salonika	21/7/17, p. 7454
Robertson, G. McM. D.S.O., M.C.	Capt. (A./Lieut.-Col.)	10th	France	24/5/18, p. 6101
		N. Staffs Regt. (attd. 10th Bn. R.D.F.)	France	21/12/17
		2nd Bn. N. Staffs Regt. (attd. 2nd Bn. Manch. Regt.)	France	28/12/18
			France	9/7/19
Shanks, W. J.	2nd-Lieut.	2nd	France	1/1/16, p. 56
Stead, T.	Pte. (Regtl. No. 17748)	—	Dardanelles	28/1/16, p. 1204
Stephens, P.	2nd-Lieut. (Temp. Capt.)	4th	France	4/1/17, p. 240
Stirke, H. R., D.S.O.	Temp. Major	9th (attd. 2nd)	France	4/1/17
			France	24/5/18, p. 6101
Synnott, F.	Sergt. (Regtl. No. 16177)	9th	France	25/5/17, p. 5166
Silcox, B.	Temp. 2nd-Lieut.	Gen. List (attd. 7th)	Salonika	25/11/17, p. 12485
Seymour, E. F. E.	Major (A./Lieut.-Col.)	10th	France	21/12/17, p. 13373
Shine, J. O. W.	Capt.	Attd. 9th	France	21/12/17, p. 13373
Shadforth, H. A.	Capt.	6th	Egypt	30/4/19, p. 5445
Shirren, A. G.	C.S.M. (Regtl. No. 15150)	1st	France	24/5/18, p. 6101
Stirling, W. F., D.S.O., M.C.	Major (R. of O.)	1st and Staff	Egypt	11/6/20, p. 6452
Shears, P. J.	Capt. (A./Major)	1st (attd. 11th Bn. Hampshire Regt.)	France	28/12/18, p. 15162
Scott, W. G.	Temp. Lieut. (A./Capt.)	4th (attd. Labour Corps)	France	9/7/19, p. 8710
Sheehy, E.	Lieut. (A./Capt.)	4th	France	9/7/19, p. 8710
Swetman, F.	Pte. (Regtl. No. 28641)	6th	France	9/7/19, p. 8710
Tredennick, J. P., D.S.O.	Major (Temp. Lieut.-Col.)	2nd (attd. 18th London Regt.)	France	1/1/16, p. 56
			France	5/7/19, p. 8500
Tarleton, G. W. B.	Lieut.	2nd	France	1/1/16, p. 56
	Capt.	—	France	21/12/17, p. 13373

Name	Rank	Battalion	Theatre	Gazette
Treacher, F., M.C.	Lieut.	2nd	France	1/1/16, p. 56
	Capt.	—	France	5/7/19, p. 8500
Thompson, P. T. L.	Capt.	6th	Dardanelles	28/1/16, p. 1204
Tobin, R. P.	Temp. Capt. (killed)	7th	Dardanelles	28/1/16, p. 1204
Thompson, A. C.	Temp. Major (Temp. Lieut.-Col.)	8th	France	4/1/17, p. 240
			France	25/5/17, p. 5166
			France	21/12/17, p. 13373
Tippet, H. C. C.	Capt.	4th and Staff	France	11/12/17, p. 12921
Tittle, D. R.	Lieut. (A./Capt.)	2nd (attd. 1st Bn. Conn. Rangers)	Egypt	5/6/19, p. 7181
Taylor, C.	Cpl. (Regtl. No. 7/24256)	6th (attd. H.Q. 10th Div.)	Egypt	12/1/20, p. 504
Wheeler, S. G. de C.	Capt.	2nd	France	17/2/15, p. 1668
	Major	—	France	28/12/18, p. 15162
Whelan, J.	Pte. (Regtl. No. 10974)	1st	Dardanelles	5/8/15, p. 7666
West, J.	Sergt. (Regtl. No. 17141)	7th	Dardanelles	28/1/16, p. 1204
Wilkinson, G. N.	Temp. Capt.	7th	Dardanelles	28/1/16, p. 1204
Williams, A. E.	Pte. (Regtl. No. 17750)	—	Dardanelles	5/5/16, p. 4519
Watson, R. M., D.S.O.	Capt.	2nd	France	17/2/15, p. 1668
			France	15/6/15, p. 5924
			France	15/6/16, p. 5923
			France	4/1/17
			France	20/12/18, p. 14935
			France	5/7/19, p. 8501
Whyte, W. H.	Temp. Major (A./Lieut.-Col.) (Lt. R. of O.)	6th	Dardanelles	28/1/16, p. 1204
			Salonika	6/12/16, p. 11939
			Salonika	28/11/17, p. 12485
Whittaker, L.	Sergt. (Regtl. No. 9709)	8th	France	4/1/17, p. 240
Watson, F. H.	Pte. (Regtl. No. 19200)	Late 4th	Home "A" List	25/1/17, p. 946
Wilson, F. A.	Temp. Lieut. (A./Capt.)	1st	France	25/5/17, p. 5166
Wallis, D.	C.S.M. (Regtl. No. 11435)	1st	France	25/5/17, p. 5166
West, M.	Pte. (Regtl. No. 8207)	1st	France	25/5/17, p. 5166
Whelan, J.	Sergt. (Regtl. No. 9210)	1st	France	21/12/17, p. 13373
Whitton, W. J.	C.S.M. (Regtl. No. 10636)	10th	—	21/12/17, p. 13373
Weldon, K. C., D.S.O.	Major (Temp. Lieut.-Col.)	2nd Bn.	France	9/7/19, p. 8710
Watson, J. W.	Sergt. (Regtl. No. 15703)	2nd	France	9/7/19, p. 8710
White, M.	Cpl. (Regtl. No. 9878)	1st	France	9/7/19, p. 8710
Wilkins, A.	Pte. (Regtl. No. 14091)	6th	France	9/7/19, p. 8710
White, H., M.C.	Sergt.-Major (Regtl. No. 9/15206)	9th	France	4/1/17, p. 240
	Temp. Lieut. (A./Capt.)	—	France	28/12/18, p. 15162
			France	9/7/19, p. 8710

P

HOME SERVICES AWARDS.

THE names of the undermentioned have been brought to the notice of the Secretary of State for War for valuable services rendered in connection with the war :—

NAME.	REGTL. NO.	RANK.	BATTALION.	"LONDON GAZETTE."
Agnew, A. E. H. ...	...	Capt.	—	13/8/18, p. 180
Annetts, A. E. ...	111126 ...	Col.-Sergt. ...	3rd ...	28/8/19, p. 400
Boyle, H.	29405 ...	W.O. Cl. II. (S.Q.M.S.)	—	4/11/18, p. 225
Briggs, G. E. ...	—	Bt. Col. (R.P.)	R.D.F.	13/8/18, p. 180
Brennan, T.	25521 ...	Pte. (A./Sergt.-Dmr.)	3rd ...	28/8/19, p. 401
Browne, J.	29406 ...	Col.-Sergt. (A./ W.O. Cl. II., S.Q.M.S.)	—	4/11/18, p. 225
Browne, J.	29406 ...	S.Q.M.S. ...	—	15/3/19, p. 273
Byrne, J.	9608 ...	R.Q.M.S. ...	—	28/8/19, p. 441
Byrne, J.	9607 ... (now 12292	Sgt. (A./R.Q.M.S.) 7th Bn. R. Ir. Regt.)	1st ...	27/3/19, p. 304
Byrne, L., D.C.M., M.M.	13/561 ...	C.Q.M.S. ...	3rd ...	28/8/19, p. 401
Caulfield, C. E. ...	10726 ...	Bandmaster ...	1st ...	27/3/19, p. 304
Collins, F. R. ...	—	Capt.	—	25/3/19, p. 278
Daly, J.	4956 ...	R.S.M. ...	3rd ...	28/8/19, p. 402
Darby, W. G. ...	29437 ...	Col.-Sergt. (A./ W.O., Cl. II., S.Q.M.S.)	—	4/11/18, p. 225
Dawson, F.	10530 ...	Sergt. (A./Col.-Sergt.)	—	25/3/19, p. 279
Dickie, T. W. ...	—	Capt.	—	24/2/17, p. 9
Doyle, O.	9897 ...	C.S.M. ...	3rd ...	28/8/19, p. 403
Drewett, J.	13518 ...	Q.M.S. ...	—	27/3/19, p. 305
Farrell, M.	13561 ...	Col.-Sergt. ...	3rd ...	28/8/19, p. 403
Fitzpatrick, F. H. ...	29404 ...	S.Q.M.S. ...	—	27/3/19, p. 305
French, D.	—	Major	—	13/8/18, p. 182
Gillespie, W. ...	16388 ...	R.S.M. ...	—	24/2/17, p. 37
Grattan-Esmonde, L.	—	Major (Temp. Lieut.-Col.)	—	24/2/17, p. 13
Greenland, H. ...	12169 ...	C.S.M. (A./ R.S.M.)	3rd ...	28/8/19, p. 404
Hoey, C. B. R. ...	—	Major	—	27/3/19, p. 298
Howis, J. E.	29440 ...	Pte. (A./Sergt.)	—	27/3/19, p. 307
Jackson, W.	20147 ...	W.O., Cl. II. (S.Q.M.S.)	Depot ...	4/11/18, p. 226
Johnson, A. M. ...	—	Major	—	13/8/18, p. 183
Keogh, E. J. L. ...	—	Major	3rd ...	28/8/19, p. 398
Lindsay, W., M.V.O.	—	Lieut.-Col. ...	—	24/2/17, p. 19
Loughran, W. ...	15805 ...	Sergt.	3rd ...	28/8/19, p. 405
McGrath, J.	5137 ...	Cpl. (A./Sergt.)	3rd ...	28/8/19, p. 406
Murphy, H. P. ...	—	Temp. Lieut. (Temp. Capt.)	—	28/8/19, p. 441
Murphy, W.	22197 ...	C.S.M. (A./ R.S.M.)	—	13/3/18, p. 150
Parry, T.	10634 ...	Cpl.	3rd ...	28/8/19, p. 406
Partridge, H. ...	5903 ...	C.S.M. ...	3rd ...	28/8/19, p. 406
Perreau, C. N. ...	—	Col., Canadian Forces	...	24/2/17, p. 49 5/7/18
Popham, F. S. ...	—	Capt. and Bt. Maj.	—	28/8/19, p. 441

NAME.		REGTL. NO.		RANK.		BATTALION.		" LONDON GAZETTE."
Reddy, T. F.	...	19183	...	Pte. (A./Sergt.)		7th	...	27/3/19, p. 309
Reid, A. A. ...	...	29426	...	Pte. (A./Sergt.)		—		4/11/18, p. 226
Reid, A. A. ...	...	29426	...	Pte. (A./Sergt.)		—		27/3/19
Shanley, P. ...	...	11308	...	Sergt. ...	...	3rd	...	28/8/19, p. 407
Smith, J. ...	...	29398	...	S.M. ...	...	—		25/3/19, p. 282
Smithwick, S. G.	...	—		Major ...	...	—		28/8/19, p. 399
Tighe, M. H.	...	—		Lieut. ...	...	3rd	...	15/3/19, p. 273
Townsend, A.	...	5736	...	C.Q.M.S.	...	3rd	...	28/8/19, p. 408

FOREIGN AWARDS, 1914–18.

NAME.	RANK.	BATTALION.	REWARDS.	" LONDON GAZETTE."
Boylan, Frank	Pte. (Regtl. No. 15155)	8/9th	Belgian Decoration Militaire	15/4/18
Brady, Thomas	Capt.	—	French Croix de Guerre avec Palme	15/12/19
Brophy, Peter	L./Cpl. (Regtl. No. 32603)	—	Russian Medal on St. Stanislas Riband	16/7/21
Brown, Frederick Elliott	Lieut.	—	French Croix de Guerre	14/7/17
Bryan, Edward	Sergt. (Regtl. No. 13197)	6th	Serbian Cross of Karageorge, 1st Class, Gold Star (with Swords)	21/4/17
Burke, John Patrick	Sergt. (Regtl. No. 12970)	9th (attd. 48th T.M. Bty.)	French Croix de Guerre	1/5/17
Byrne, John	Pte. (Regtl. No. 9590)	2nd	Montenegrin Medal for Merit (silver)	9/3/17
Byrne, Lawrence	C.Q.M.S. (Regtl. No. 8901)	1st	French Croix de Guerre	17/12/17
Campbell, Charles Wilson	C.Q.M.S. (Regtl. No. 24700)	10th	Belgian Croix de Guerre	12/7/18
Connolly, Joseph	L./Sergt. (Regtl. No. 9392)	2nd	French Croix de Guerre	14/7/17
Cooke, William	Sergt. (Regtl. No. 8672)	2nd	Russian Medal of St. George, 2nd Class	25/8/15
Corballis, Edward Roux Littledale	Capt.	R.D.F and R.F.C.	French Legion d'Honneur (Chevalier)	14/7/17
Cory, George Norton	Major and Bt. Lieut.-Col.	R.D.F.	Russian Order of St. Anne, 3rd Class (with Sword)	15/2/17
	Major-Gen.	R.D.F.	Greek Order of the Redeemer (Grand Commander)	10/10/18
	Bt. Col. (Temp. Major-Gen.)	General Staff	Serbian White Eagle, Class III (with Swords)	7/6/19
	Major-Gen.	M.G.G.S., G.H.Q.	Greek M.C., 2nd Class	21/7/19
	Major-Gen	General Staff	French Croix de Guerre avec Palme	21/7/19
	Major-Gen.	General Staff	French Legion of Honour (Commandeur)	21/8/19
Courtney, John	Cpl. (L./Sergt.) (Regtl. No. 22493)	9th	Russian Cross of St. George, 4th Class	15/2/17
Cox, Patrick Godfred Ashley	Capt. and Bt. Major (Temp. Lieut.-Col.)	6th	French Croix d'Officier, Légion d'Honneur	21/4/17
Cummins, William	Sergt. (Regtl. No. 6603)	1st	Serbian Cross of Karageorge (with Swords), 4th Class	15/2/17
Darling, Sydney George, M.C.	2nd-Lieut. (A./Capt.)	1st (S.B.) attd. 1st	French Croix de Guerre avec Etoile en argent	19/6/19
Devoy, Joseph	Sergt. (Regtl. No. 10335)	1st	French Medaille Militaire	24/2/18
Dillon, John	Pte. (Regtl. No. 40436)	1st	Belgian Croix de Guerre	12/7/16

Name	Rank	Battalion	Honour	Date
Doogan, Michael Joseph	A./C.S.M. (Regtl. No. 10902)	R.D.F., attd. Army Gym. Staff	Belgian Croix de Guerre	4/9/19
Elston, William Thomas	Cpl. (A./L./Sergt.) (Regtl. No. 15346)	2nd	Belgian Croix de Guerre	12/7/18
Falkiner, Frederick	L./Sergt. (Regtl. No. 7/14166) ; 2nd-Lieut. R. Ir. Rifles	7th	Italian Bronze Medal for Military Valour	31/8/17
Gavin, Peter	Pte. (A./L./Cpl.) (Regtl. No. 6/11937)	6th	Serbian Silver Medal	15/2/17
Gibson, George, M.M.	Cpl. (Regtl. No. 16531)	1st	French Croix de Guerre avec Etoile en bronze	19/6/19
Glegg, James Dowie, M.C.	Capt.	—	Belgian Croix de Guerre	4/9/19
Gorry, Joseph	Cpl. (A./L./Sergt.) (Regtl. No. 24825)	8/9th	Belgian Croix de Guerre	12/7/18
Grove, James Robert Wood	Bt. Major	—	Russian Order of St. Anne, 3rd Class, with Swords and Bow	16/7/21
Hall, R. S.	C.S.M. (Regtl. No. 5039)	2nd	French Medaille Militaire	5/11/14
Halligan, Joseph Thomas, O.B.E.	Capt. (Temp. Major)	—	French Croix de Guerre avec Palme	8/3/20
Harcourt, Harry Gladwyn, D.S.O., M.C.	Lieut.	R.D.F. (attd. M.G.C.)	Russian Order of St. Stanislas, 2nd Class (with Swords)	16/7/21
Harvey, Charles Dacre	Temp. Capt.	7th	Egyptian Order of the Nile, 4th Class	9/11/18
Hawes, Cecil Ernest	Temp. Lieut. (Temp. Capt.)	—	Serbian White Eagle, 5th Class	15/2/17
Hayes, Robert	Pte. (Regtl. No. 5780)	1st	French Croix de Guerre	24/2/16
Healy, Patrick	Pte. (Regtl. No. 21450)	8/9th	Belgian Decoration Militaire	15/4/18
Hoey, Charles Bayly Robert	Major	—	Order of the Star of Roumania (Officer)	20/9/19
Holmes, Geoffrey Gordon	Temp. Lieut.	1st	French Croix de Guerre avec Etoile en bronze	19/6/19
Higginson, H. W., C.B., D.S.O.	Major and Bt. Col. (Temp. Major-Gen.)	Staff	Order of the Star of Roumania (Commandeur)	-/-/19
			French Légion d'Honneur (Officier)	-/-/19
Irwin, Thomas	L./Cpl. (Regtl. No. 5728)	Spec. Res. (attd. 1st)	French Croix de Guerre	24/2/16
Jenkins, Cuthbert Esmond	Temp. Lieut. (A./Capt.)	—	French Croix de Guerre avec Palme	17/8/18
Kavanagh, Martin	Sergt. (Regtl. No. 13431)	9th	French Croix de Guerre	14/7/17
Kendrick, Edward Holt, D.S.O.	Bt. Major (Temp. Lieut.-Col.)	R.D.F. & M.G.C.	French Légion d'Honneur (Chevalier)	22/11/18
			French Légion d'Honneur (Officier)	21/8/19

NAME.	RANK.	BATTALION.	REWARDS.	"LONDON GAZETTE."
Leahy, Thomas Joseph	Lieut.	2nd	French Croix de Chevalier, Légion d'Honneur	3/11/14
Lynch, Christopher	L./Cpl. (Regtl. No. 8542)	1st	French Croix de Guerre avec Etoile en Vermeil	14/7/17
McLoughlin, Michael J.	Sergt. (Regtl. No. 25245)	10th	Belgian Croix de Guerre	12/7/18
McNeely, Joseph	Sergt. (Regtl. No. 23998)	1st	Belgian Croix de Guerre	4/9/19
Martin, Charles Andrew	Temp. Lieut.	Service	Serbian Order of the White Eagle, 5th Class	9/3/17
Metcalfe, Maurice Owen	Cpl. (A./Sergt.) (Regtl. No. 14289)	10th	French Croix de Guerre	10/10/18
Moore, Patrick	Pte. (Regtl. No. 8272)	1st	Serbian Silver Medal	15/2/17
Neill, Charles	Temp. Lieut.	1st	Belgian Croix de Guerre	4/9/19
O'Carrol, Frederick W. J.	Capt.	5th	French Croix de Guerre	21/7/19
Ockenden, James, V.C.	Sergt. (Regtl. No. 10605)	1st	Belgian Croix de Guerre	12/7/18
O'Dowda, James Wilton	Lieut.-Col. and Bt. Col. (Temp. Brig.-Gen.)	—	Serbian White Eagle, 3rd Class	15/2/17
O'Leary, James	Sergt. (Regtl. No. 10310)	1st	Belgian Decoration Militaire avec Croix de Guerre	26/11/19
Plunkett, Joseph Frederick, D.S.O., M.C., D.C.M.	Capt. (Temp. Lieut.-Col.)	R.D.F. (attd. 13th Bn. R. Innis. Fus.)	French Croix de Guerre avec Etoile en Vermeil	19/6/19
Ray, A. W.	(Regtl. No. 6704)	2nd	French Medaille Militaire	5/11/14
Roche, Michael	Cpl. (A./Sergt.) (Regtl. No. 10570)	2nd	Roumanian Medaille Barbatie si Credenta (3rd Class)	20/9/19
Romer, Cecil Francis, C.B., C.M.G., A.D.C.	Lieut.-Col. and Bt. Col., Major-Gen.	General Staff	Russian Order of St. Anne, 2nd Class	15/2/17
			French Légion d'Honneur, Croix d'Officier	2/6/17
Sinnott, Lawrence	C.S.M. (Regtl. No. 6/10485)	6th	Russian Cross of St. George, 3rd Class	15/2/17
Stirke, Henry Richard, D.S.O.	Temp. Major	—	Belgian Croix de Guerre	12/7/18
Stirling, Walter Francis, D.S.O., M.C.	Major (Temp. Lieut.-Col.)	R.D.F., Retired Pay (R. of O.)	Hedjaz, The Nahda Order, 2nd Class	8/3/20
Stokes, James	Pte. (Regtl. No. 9150)	2nd	French Medaille Militaire	14/7/17
Swifte, Latham Coddington	Major	5th	Italian Croce di Guerra	21/8/19
Tansey, Patrick	L./Cpl. (Regtl. No. 6/13135)	6th	Serbian Gold Medal	15/2/17
Tarleton, Gerald Waldon Browne, M.C.	Lieut.	R.D.F. (attd. Service Bn.)	Serbian Order of the White Eagle, 5th Class	21/4/17
Waine, Patrick, D.C.M.	C.S.M. (Regtl. No. 11167)	1st	Belgian Croix de Guerre	12/7/18
Weldon, Kenneth Charles	Major (Temp. Lieut.-Col.)	Attd. 7/8th R. Ir. Fus.	French Légion d'Honneur (Officier)	14/7/17
Whyte, William Henry	Lieut. (Temp. Major)	R. of O., R.D.F. (attd. R. Ir. Fus.)	Serbian Order of the White Eagle, 4th Class	21/4/17

HONOURS SELECTED FOR REGIMENTAL COLOURS AND ARMY LIST.

The following is a list of Honours in connection with the Great War to be placed on the Regimental Colours, together with additional Honours for insertion in the Army List, as selected by the Regimental Committee composed as under, and in accordance with instructions contained in War Office letter No. 20/Gen., No. 5000 (Q.M.G. 7) of February 15th, 1923, Army Order 338, dated September 4th, 1922, and Army Council Instruction No. 458 of September 8th, 1922 :—

The Committee met in London on January 16th and March 12th, 1923, and Battalions were represented as follows :—

Chairman—Major-General C. D. Cooper, C.B., Colonel Royal Dublin Fusiliers.

Secretary—Lieut.-Colonel C. N. Perreau, C.M.G., late Commanding 1st Bn. Royal Dublin Fusiliers.

Representing :—

1st Battalion—Bt. Major T. J. Carroll-Leahy, D.S.O., M.C.

2nd Battalion—Col.-Comdt. H. W. Higginson, C.B., D.S.O., A.D.C.

      Bt. Lieut.-Colonel K. C. Weldon, D.S.O.
      Major J. Burke, D.S.O., M.C., D.C.M.

5th Battalion—Lieut.-Colonel F. H. Macnamara.

6th Battalion—Capt. J. Esmonde, M.C.

8th Battalion—Lieut.-Colonel Sir E. H. C. P. Bellingham, Bt., C.M.G., D.S.O.

9th Battalion—Lieut.-Col. J. P. Hunt, C.M.G., D.S.O., D.C.M.

10th Battalion—Lieut.-Col. E. F. Seymour, D.S.O., O.B.E.

NOTE.—Representatives of the 3rd, 4th, and 7th Battalions were unavoidably absent, but sent letters in connection with the war services of their Battalions.

LIST OF HONOURS FOR THE COLOURS.

MONS.	SOMME, 1916, 1918.
MARNE.	CAMBRAI, 1917, 1918.
YPRES, 1915, 1917.	PALESTINE, 1917, 1918.
GALLIPOLI, 1915, 1916.	HINDENBURG LINE.
MACEDONIA, 1915, 1916, 1917.	SELLE.

ADDITIONAL HONOURS FOR INSERTION IN ARMY LIST.

Operations.	Battles. (For Honours in Army List.)	Battalion of Regiment on behalf of whom Honour was claimed.	Brigade, Division, or Formation serving with.
Retreat from Mons : Aug. 23rd–Sept. 5th, 1914	1. Le Cateau	2nd Battalion ...	10th Bde., 4th Div.
Advance to the Aisne : Sept. 6th–Oct. 1st, 1914	2. Retreat from Mons ... ... 3. Aisne, 1914	,, ,, ... ,, ,, ...	,, ,, ,, ,, ,, ,,
Operations in Flanders, 1914 : Oct. 10th–Nov. 22nd	4. Armentières	,, ,, ...	,, ,, ,,
Summer Operations, 1915 : Mar.–Oct. ...	5. St. Julien ... ... 6. Frezenberg	,, ,, ... ,, ,, ...	,, ,, ,, Attached 84th Brigade.
Operations on the Somme : July 1st–Nov. 18th, 1916	7. Bellewaerde ... ... 8. Albert ... ... 9. Delville Wood ... ... 10. Guillemont	,, ,, ... ,, ,, ... 1st Battalion ... 2nd Battalion ... 8th Battalion ... 9th Battalion ...	10th Bde., 4th Div. ,, ,, ,, 86th Bde., 29th Div. 10th Bde., 4th Div. 48th Bde., 16th Div. ,, ,, ,,
	11. Ginchy	8th Battalion ... 9th Battalion ...	,, ,, ,, ,, ,, ,,
	12. Morval ... ... 13. Transloy ... ... 14. Ancre Heights	2nd Battalion ... ,, ,, ... 10th Battalion ...	10th Bde., 4th Div. ,, ,, ,, 189th Bde., 63rd Naval Division.
	15. Ancre, 1916	,, ,, ...	189th Bde., 63rd Naval Division.
The Arras Offensive : April 9th–May 15th, 1917	16. Arras, 1917 (Scarpe battles) ... 17. Scarpe, 1917	1st Battalion ... 10th Battalion ...	86th Bde., 29th Div. 189th Bde., 63rd Naval Division.
Flanking operations round Bullecourt : April 11th–June 16th	18. Bullecourt	2nd Battalion ... 8th Battalion ... 9th Battalion ...	48th Bde., 16th Div. ,, ,, ,, ,, ,, ,,
Flanders Offensive : June 7th–Nov. 10th, 1917	19. Messines, 1917	2nd Battalion ... 8th Battalion ... 9th Battalion ...	,, ,, ,, ,, ,, ,, ,, ,, ,,

Flanders Offensive : June 7th–Nov. 10th, 1917 (*contd.*)	20. Langemarck, 1917	2nd Battalion ...	48th Bde., 16th Div.
		8th Battalion ...	,, ,, ,,
		9th Battalion ...	,, ,, ,,
The Offensive in Picardy : Mar. 21st–April 5th, 1918	21. St. Quentin	1st Battalion ...	,, ,, ,,
		2nd Battalion ...	,, ,, ,,
	22. Bapaume, 1918	8th Battalion ...	,, ,, ,,
		9th Battalion ...	,, ,, ,,
The Breaking of the Hindenburg Line : Aug. 26th–Oct. 12th, 1918	23. St. Quentin Canal	2nd Battalion ...	148th Bde., 50th Div.
	24. Beaurevoir	...	198th Bde., 66th Div.
		6th Battalion ...	
	25. Cambrai, 1918	2nd Battalion ...	148th Bde., 50th Div.
Final advance	26. Sambre	...	,, ,, ,,
		6th Battalion ...	,, ,, ,,

MACEDONIA.

Retreat from Serbia on Salonika : Dec., 1915	27. Kosturino	6th Battalion ...	30th Bde., 10th Div.
		7th Battalion ...	,, ,, ,,
Operations in Struma Valley : 1916–1918 ...	28. Struma	,, ,, ...	,, ,, ,,

DARDANELLES.

Helles Operations : April 25th, 1915–Jan. 8th, 1916	29. Landing at Helles	1st Battalion ...	86th Bde., 29th Div.
	30. Krithia	,, ,, ...	,, ,, ,,
Anzac and Suvla Operations : April 25th–Dec. 20th, 1915	31. Suvla	,, ,, ...	,, ,, ,,
	32. Sari Bair	1st Battalion ...	86th Bde., 29th Div.
		6th Battalion ...	30th Bde., 10th Div.
	33. Landing at Suvla	7th Battalion ...	,, ,, ,,

EGYPT AND PALESTINE.

Second Offensive : Oct. 27th–Nov. 16th, 1917	34. Gaza	6th Battalion ...	30th Bde., 10th Div.
		7th Battalion ...	,, ,, ,,
	35. El Mughar	,, ,, ...	,, ,, ,,
Jerusalem Operations : Nov. 17th–Dec. 30th, 1917	36. Nebi Samwil	,, ,, ...	,, ,, ,,
	37. Capture of Jerusalem	,, ,, ...	,, ,, ,,

[*Appendix Five*

DESCRIPTION OF MEMORIALS erected to the Memory of Officers, W.Os., N.C.Os. and Men of The Royal Dublin Fusiliers who laid down their lives for King and Country in the Great War, 1914–1918.

IN WESTMINSTER CATHEDRAL.

With the approval and co-operation of the surviving Officers and Men of the Irish Regiments a committee was formed shortly after the cessation of hostilities of the Great War (1914–1918), for the purpose of establishing a permanent Memorial to the Officers, Non-commissioned Officers, and Men of Irish Regiments killed in that war, and it was proposed that this Memorial should take the form of a Chapel dedicated in Westminster Cathedral, with books containing the names of those who lost their lives.

Cardinal Bourne offered St. Patrick's Chapel in the Cathedral for this purpose, and, although the original idea was to decorate and complete the entire Chapel, owing to lack of funds, modifications had to be made. Each Regiment which has subscribed is to be represented by a panel in the Chapel, and the books of names are to be kept in a suitable receptacle.

There can be no more fitting place for such a record than a Cathedral in the heart of the British Empire. The Chapel will be a lasting Memorial to those who so gallantly and grandly upheld the honour of the Irish Regiments, and it is hoped to make it a shrine worthy of the supreme sacrifice.

IN THE CHAPEL OF THE ROYAL MILITARY COLLEGE, SANDHURST.

A Memorial to Officers of the Royal Dublin Fusiliers, who were Cadets at the Royal Military College.

The Memorial will consist of a White Marble Slab with the names of the Officers as under, surrounded by a frame of carved oak, and with the crest of the Regiment coloured and carved in the oak at the top.

Inscription on R.M.C. Memorial.

Roll of Officers,
Cadets at the R.M.C.,
who served in
The Royal Dublin Fusiliers
and who
Died for King and Country
in
The Great War of 1914–1918.

1914

Lieut.-Colonel P. Maclear   Captain H. O. Davis

1915

Capt. D. V. F. Anderson   2/Lieut. R. A. F. S. King
Lieut. R. V. C. Corbet   2/Lieut. M. C. N. R. Young
Lieut. R. de Lusignan   Capt. B. Maclear
Capt. G. M. Dunlop   2/Lieut. H. D'E. Head
Bt. Major T. H. C. Frankland   Capt. A. A. C. Taylor
Capt. F. N. Le Mesurier   Lieut. L. C. Boustead
2/Lieut. B. McGuire   Capt. H. M. Floyd
Lieut. R. Bernard   Lieut. H. D. O'Hara, D.S.O.
Major E. Fetherstonhaugh   2/Lieut. J. A. H. Taylor
2/Lieut. M. O'C. Cuffey   2/Lieut. H. R. T. Hackett.

1916

Capt. E. R. L. Maunsell   Major D. E. Wilson
2/Lieut. W. H. A. Damiano   2/Lieut. H. J. Lemass
2/Lieut. J. A. H. Helby   Lieut. H. G. Killingley
Lieut. S. V. C. Jones

1917

2/Lieut. G. F. Gradwell
Lieut. H. D. K. George
Capt. H. L. Ridley, M.C.

Lieut. F. Dowling
Capt. J. O. W. Shine
2/Lieut. G. S. Falkiner

1918

2/Lieut. E. H. Robertson
2/Lieut. M. J. Macnamara
2/Lieut. R. G. Hunter

Lieut. G. P. N. Thompson
Lieut. W. Pedlow, M.C.
Bt. Lieut.-Col. A. W. Gordon

1919

Lieut.-Col. H. R. Beddoes

This Memorial is erected by their Comrades in proud recognition of their gallantry and supreme sacrifice, and of all other Officers, Warrant Officers, Non-commissioned Officers and Men of The Royal Dublin Fusiliers who gave their lives in the Great War, 1914–1919.

" Spectamur Agendo."

[*Appendix Six*

WHAT OUR WOMEN DID FOR THE REGIMENT
During the Great War.

The work for the men of the Royal Dublin Fusiliers was inaugurated by the Committee of the Dublin Women's Unionist Club at a meeting held in August, 1914, at the offices of the Club at No. 10, Leinster Street, Dublin, and commenced with the collection and dispatch of a large quantity of newspapers, magazines and books to the battalions of the Regiment serving with the Expeditionary Force in France and Flanders, until the Camp Libraries made unnecessary any individual provision of literature for the troops.

The very urgent appeal made by an Irish woman to the Committee of the Club early in 1915 for food to be sent to those of the Royal Dublin Fusiliers who were prisoners of war at Limburg in Germany, caused the Committee to concentrate their energies on this work, until ever-increasing numbers and the regulations published by Government necessitated larger measures of relief. One of the workers then proposed the co-ordination of work of all kinds for the Regiment under a Central Advisory Committee for the Regimental Area of the City and County of Dublin and the Counties of Wicklow, Carlow and Kildare, and this Advisory Committee then became " The Dublin County Association for the Administration of Voluntary Work," and " The Regimental Care Committee for Prisoners of War " as approved by the War Office.

The following were some of the Branches of the Central Advisory Committee :—

Women's Branch, Royal Dublin Fusiliers Clothing Committee, 102, Grafton Street, Dublin. (Clothing for men at the Front and some branches of the Prisoners of War Committee.) Chairman : Lady Moore. Hon. Secretary : Miss Dickson.

The Dublin Committee, Prisoners of War, 65, Merrion Square. (Food and Clothing.) President : Lady Arnott. Secretary : Miss Rosalie Hamilton. Hon. Treasurer : Miss Croker.

The County Kildare Committee, Courthouse, Naas. (Food for Prisoners of War.) President : The Countess of Mayo.

Dundrum, County Dublin, Prisoners of War Committee. (Food.) President : Mrs. Arthur Goff.

Bray, County Wicklow, Prisoners of War Committee. (Food.) President : Miss Jameson.

The Bonded Store, under the direct control of the Customs and Excise officials, for all branches, was in charge of the Dublin Committee, 65, Merrion Square, and from this branch alone over 50,000 parcels were packed by a small band of no more than 10 helpers.

The final effort of the four Counties was a Mammoth Auction of valuable gifts, held in Dublin in October, 1918, just before the Armistice, and which realised £15,000. The balance of this sum after paying off all debts amounted to £11,000 and this was handed over by the Master of the Rolls to the Trustees for the benefit of those men of the Royal Dublin Fusilier battalions who had been prisoners of war, or their dependents, and afterwards for the benefit and advancement of those of the Regiment who had served outside the United Kingdom during the war and who had not been prisoners of war.

The following were appointed trustees : Lady Arnott, D.B.E., Right Hon. L. Jonathan Hogg, and Mr. Andrew Jameson, D.L.

Perhaps the words spoken by a Woman to the first batch of repatriated Prisoners of War on arrival in Ireland from Germany, voice all that was felt and done for our soldiers during the tragic years of the Great War : "Soldiers, Welcome Home ! All have thought of you, prayed for you, and now we bid you the warmest and heartiest of welcomes. We are thankful to see you back in Ireland ; we rejoice that your sufferings are over and that you are coming back to those who love and care for you. We trust you will have many and happy years among those who love you, and among your fellow-countrymen who are proud of you."

The Regimental History Committee have thought that the War Record would be incomplete did it not contain this very brief account of all that Irish women did for the Regiment.

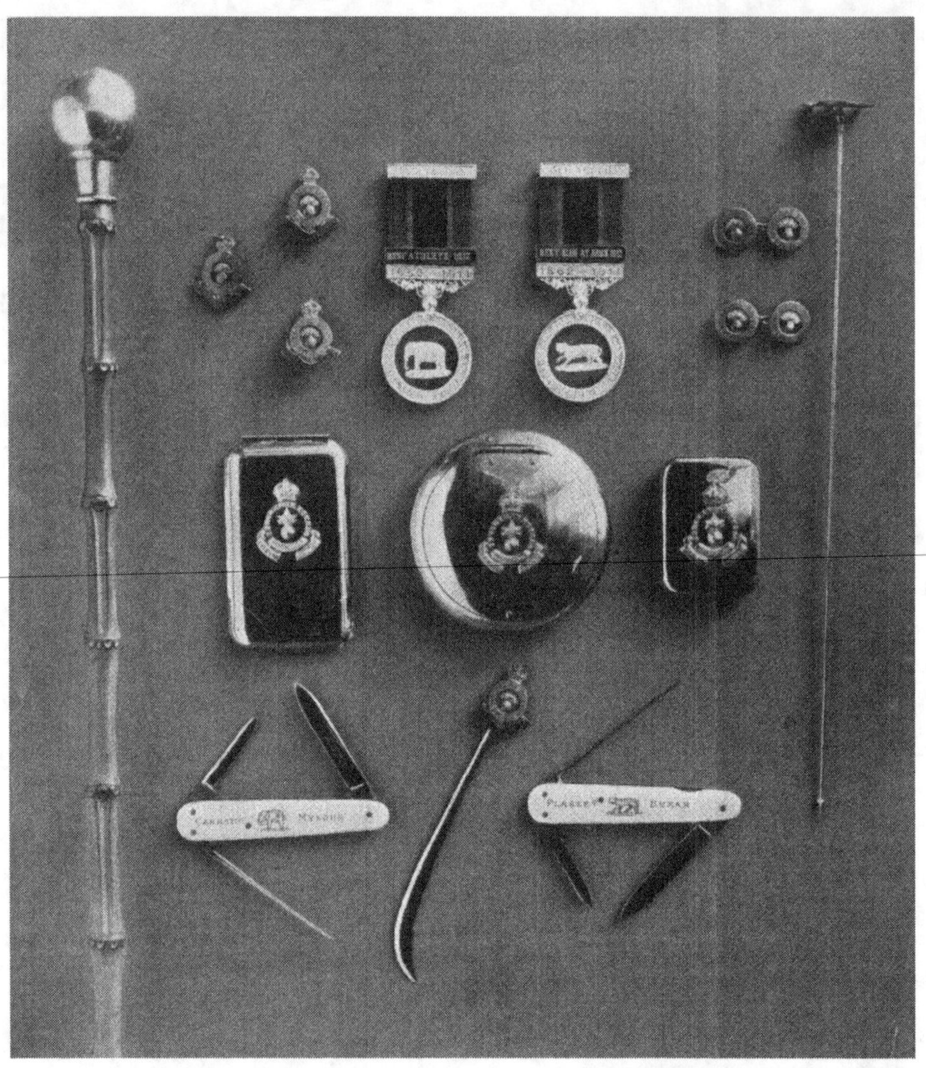

REGIMENTAL SOUVENIRS.

OFFICERS ON THE STRENGTH OF THE 2ND BATTALION, ROYAL DUBLIN FUSILIERS, JULY 31st, 1922 (date of Disbandment), at Bordon.

Lieut.-Colonel G. S. Higginson (*Commanding*).
Major and Brevet Lieut.-Colonel J. McD. Haskard, C.M.G., D.S.O.
Major S. G. Smithwick, O.B.E.
Major D. French.
Captain J. M. Mood, O.B.E., M.C.
Captain C. G. Carruthers, M.C.
Captain G. W. B. Tarleton, M.C.
Captain J. D. Glegg, M.C.
Captain C. W. Maffett.
Captain T. Brady.
Lieutenant C. Matson, M.C.
Lieutenant W. H. Stitt, D.S.O., M.C. (*Adjutant*).
Lieutenant C. M. Craig-McFeely, D.S.O., M.C.
Lieutenant H. G. Aylmer.
Lieutenant W. H. Hynes.
Lieutenant H. B. Harrison, M.C.
Lieutenant L. A. Lawrence (*Depot*).
Lieutenant D. C. A. Shepard.
Lieutenant J. J. Dolan, M.C.
Lieutenant T. E. Flewett.
Lieutenant T. A. H. Chadwick.
Lieutenant D. S. Norman.
Lieutenant M. A. Condron.
Lieutenant M. H. FitzGerald.
Lieutenant G. D. B. Russell.

[Appendix Eight

A LIST OF OFFICERS OF THE 1st AND 2nd BATTALIONS THE ROYAL DUBLIN FUSILIERS SERVING ON DISBANDMENT, showing Units to which they were transferred, etc.

LIEUT.-COLONELS.

Perreau, C. N.	
Higginson, G. S.	Retires. 1/8/22.

MAJORS.

Haskard, J. McD.(Bt.Lt.-Col.)	East Yorkshire Regt. 5/8/22.
Jeffreys, R. G. B.	North Staffordshire Regt. 5/8/22.
Wheeler, S. G. de C. ...	Welch Regt. 9/9/22.
Smithwick, S. G.	Duke of Cornwall's Light Infantry. 19/8/22.
Weldon, K. C. (Bt. Lt.-Col.)...	Sherwood Foresters. 2/8/22.
Tredennick, J. P.	Bedfordshire and Hertfordshire Regt. 29/7/22.
Knox, R. F. B.	King's Regt. 19/8/22.
Robinson, J. P. B.	Royal Berkshire Regt. 12/8/22.
Hoey, C. B. R.	Retires on Retired Pay. 15/7/22.
Crozier, H. C.	King's Regt. (Liverpool)
French, D.	Retires on Retired Pay. 27/8/22.

CAPTAINS.

Mood, J. M.	East Yorkshire Regt. 26/8/22.
Watson, R. M.	Lancashire Fusiliers. 29/7/22.
Grove, J. R. W.	Bedfordshire and Hertfordshire Regt. 6/9/22.
Dobbs, J. F. K.	King's Regt. 6/6/22.
Carroll-Leahy, T.	Northumberland Fusiliers. 29/7/22.
Lanigan-O'Keeffe, F. S. ...	Royal Welch Fusiliers. 7/10/22.
Braddell, W. H.	Northumberland Fusiliers. 29/7/22.
Shadforth, H. A.	Royal Scots Fusiliers. 2/9/22.
Massy-Westropp, R. F. H.	Lancashire Fusiliers. 2/9/22.
Carruthers, C. G.	Border Regt. 12/8/22.
Shears, P. J.	Border Regt. 12/8/22.
Kendrick, E. H.	Tank Corps.

Tarleton, G. W. B. ... ... Border Regt. (India). 2/8/22.
Glegg, J. D. ... ... East Yorkshire Regt. 26/8/22.
Elsworthy, A. L. ... ... Royal Scots Fusiliers. 16/9/22.
Maffett, C. W. ... ... Black Watch. 12/8/22.
Treacher, F. ... ... South Staffordshire Regt. 16/9/22.
Gillett, N. C. ... ... Tank Corps.
Walters, J. P. ... ... Duke of Cornwall's Light Infantry. 7/10/22.
Caldbeck, W. E. ... ... Bedfordshire and Hertfordshire Regt. 2/9/22.
Brady, T. ... ... Retires, Receiving a Gratuity. 23/7/22.

LIEUTENANTS.

Tittle, D. R. ... ... King's Own Regt. 30/8/22.
Byrne, L. C. ... ... Royal Scots Fusiliers. 2/9/22.
O'Morchoe, K. G. ... ... Gordon Highlanders. 16/9/22.
Moffat, J. B. .. .. East Yorkshire Regt. 9/9/22.
Arnold, W. J. ... ... East Lancashire Regt. 20/9/22.
Matson, C. ... ... Gloucestershire Regt. 4/10/22.
Douglas, R. G. ... ... Gloucestershire Regt. 4/10/22.
Stitt, W. H. ... ... Welch Regt. 9/9/22.
Craig-McFeely, C. M. ... Border Regt. 26/7/22.
Power, F. T. A. ... ... Royal Corps of Signals, Welch Regt.
Williamson, U. A. F. ... Tank Corps.
Gamble, C. A. ... ... Royal Army Service Corps (Sherwood Foresters). 20/9/22.
Aylmer, H. G. ... ... Essex Regt. 20/9/22.
Hynes, W. H. ... ... Gloucestershire Regt. 14/10/22.
Carroll, S. J. M. ... ... Cameronians (Scottish Rifles). 9/9/22.
Harcourt, H. G. ... ... South Wales Borderers. 9/9/22.
Mulholland, G. F. ... ... Royal Warwickshire Regt. 20/9/22.
Harrison, H. B. ... ... Royal Welch Fusiliers. 26/8/22.
Lawrence, L. A. ... ... West Yorkshire Regt. 12/8/22.
Shepard, D. C. A. ... ... Border Regt.
Dolan, J. J. ... ... Manchester Regt.
Flewett, T. E. ... ... Border Regt.
Chadwick, T. A. H. ... ... Norfolk Regt. 5/8/22.
Norman, D. S. ... ... East Yorkshire Regt. 26/8/22.
Condron, M. A. ... ... Retires, Receiving a Gratuity. 20/7/22.
Watson, L. R. C. ... ... Royal Corps of Signals, Northumberland Fusiliers. Seconded. 4/8/22.
FitzGerald, M. H. ... ... East Yorkshire Regt. 26/8/22. Seconded for Service Royal Air Force. 16/8/22.

Q

Russell, G. D. R. ... ... Prince of Wales's Volunteers. 29/7/22.
Harrison, J. H. S. ... ... West Yorkshire Regt. 29/7/22. King's African Rifles. 1/9/22.
Renny, P. L. G. ... ... South Staffordshire Regt. 2/8/22.
Eassie, W. J. F. ... ... Loyal North Lancashire Regt. 29/7/22.

SECOND-LIEUTENANTS.

Hegan, K. E. ... ... Loyal North Lancashire Regt. 29/7/22.
Rooth, R. G. ... ... West Yorkshire Regt. 12/8/22.
Tarleton, B. ... ... Northumberland Fusiliers. 5/8/22.
Houghton, H. W. H. ... Sherwood Foresters. 26/8/22.

QUARTERMASTERS.

Major Holloway, L. ... ... York and Lancaster Regt.
Capt. Williams, A. R. ... Tank Corps. 2/10/22.
Capt. Dowling, J. E. ... Black Watch. 25/9/22.

COPY OF ARMY LIST FOR JULY, 1922
(THE LAST BEFORE DISBANDMENT)
SHOWING ROLL OF OFFICERS OF THE ROYAL DUBLIN FUSILIERS
1ST AND 2ND BATTALIONS.

1545	**1546**	**1547**	**1548**

THE ROYAL DUBLIN FUSILIERS.
[*To be Disbanded.*]
[102]
No. 11 District.

The Royal Tiger, superscribed "Plassey," "Buxar," and with motto, '*Spectamur Agendo*' underneath. The Elephant, superscribed, "Carnatic," "Mysore." "Arcot," "Condore," "Wandiwash," "Pondicherry," "Guzerat," "Sholinghur," "Nundy Droog," "Amboyna," "Ternate," "Banda," "Seringapatam," "Kirkee," "Maheidpoor," "Beni Boo Alli," "Ava," "Aden," "Mooltan," "Goojerat," "Punjaub," "Pegu," "Lucknow," "Relief of Ladysmith," "South Africa, 1899-1902,"

Agents—Mr C. R. McGrigor, Bt. & Co.

Regular and Militia Battalions.
Uniform—Scarlet. *Facings*—Blue.

1st Bn. (102nd Foot)	Bordon.	3rd Bn. (Kildare Mil.)	Naas.
2nd „ (103rd „)	Bordon.	4th „ (R. Dublin City Mil.)	Dublin.
Depôt	Bordon.	5th „ (Dublin County Mil.) ..	Dublin
		Record Office .. Hamilton.	

Colonel-in-Chief	Field-Marshal H.R.H. *The Duke of* Connaught and Strathearn, K.G., K.T., K.P., G.C.B., G.C.S.I., G.C.M.G., G.C.I.E., G.C.V.O., G.B.E., Col. Gren Gds. and R.A.S.C., and Col. in Chief The Inniskillings, H.L.I., Rifle Bde. and R.A.M.C., *Personal A.D.C. to the King* - - 7Nov 03	
Colonel	Cooper, Hon. Maj.-Gen. C. D., C.B., ret. pay - 13Mar.16	
Officer Commanding Depôt	Weldon, Maj. (bt. lt.-col.) K. C., D.S.O., R. D. Fus. .. 28Dec.21	

1st and 2nd Battalions. (Regular.)

Lt.-Colonel.
1Perreau, O. N., C.M.G. 26May19
2Higginson, G. S. 24Sept.21

Majors.
2Haskard, J. McD., C.M.G., D.S.O., p.s.c. [l] 28Apr.15, bt. lt.-col. 1Jan.17
Jeffreys, R. G. B., D.S.O. 1Sept.15
1Wheeler, S. G. de C., O.B.E. 1Sept.15
2Smithwick, S. G., O.B.E. 1Sept.15
d. 1Weldon, K. C., D.S.O. 1Sept.15, bt. lt.-col. 3June19
1Tredennick, J.F., D.S.O., O.B.E., (s.c.) 1Sept.15
e.a.Knox, R. F. B. 1Sept.15
s. Robinson, J. F. B., C.M.G., D.S.O., p.s.c. 1Sept.15, bt. lt.-col. 24June23
(1) Hoey, C. B. R. 13Jan.17
Crosier, H. C., M.C. (Supt. Phys. Trng.) 13Jan.17
2French, D. 13Jan.17

Captains.
2Mood, J. M., O.B.E., M.C. 2Aug.13
1Watson, R. M., D.S.O. 5Aug.16, bt. maj. 1Jan.17
t. Grove, J. R. W. 7Sept.16, bt. maj. 3June19
e.a.Dobbs, J. F. K., M.C. 14Dec.14
1Carroll-Leahy, T. J., D.S.O., M.C. (S.C.) *Adjt.* 28Apr.15, bt. maj. 3June19
Lanigan-O'Keeffe, F. S., M.B.E. (Spec. Empld.) 28Apr.15
1Braddell, W. H. 10Oct.15
e.a.Shadforth, H. A., O.B.E., M.C. 10Oct.15
s. Massy-Westropp, R. F. H., M.C. 10Oct.15
2Carruthers, C. G., M.C. 26Jan.16
(1) Shears, P. J. 15Aug.16
t.c. Kendrick, E. H., D.S.O. 15Oct.16, bt. maj. 3June19
2Tarleton, G. W. B., M.C. 1Jan.17
2Glegg, J. D., M.C. 1Jan.17
Elsworthy, A. L., M.B.E. (l) 1Jan.17

Captains—contd.
2Maffett, C. W. 1Jan.17
s. Treacher, F., M.C. 1Jan.17, bt. maj. 3June19
t.c. Gillett, N. C. 30Apr.17
t.c Walters, J. P. 14Sept.17
m. Caldbeck, W. E. 9Sept.30 / 22June18
(1) 2Brady, T. 22June18

Lieutenants
1Tittle, D. R. 30Aug.15
c.o.Byrne, L. C., D.S.O., M.C. 26Jan.16
c.o. O'Morchoe, K. G. 6June16
d. 1Moffat, J. B. 6June16
c.o. Arnold, W. J. 30June16
2Matson, C., M.C. 24June16
c.a.Douglas, R. G. 24July16
2Stitt, W. H., D.S.O., M.C., *Adjt.* 24July16
2Craig-McFeely, C. M., D.S.O., M.C. 24July16
c.s. Power, F.T.A., M.C. 30Dec.16
t.c. Williamson, U A. F. 1Jan.17
1Gamble, C. A. 1Jan.17
2Aylmer, H. G. 1Jan.17
2Hynes, W. H. 1Jan.17
t.c. Carroll, S. J. M.14Nov.17
t.v.Harcourt, H. G., D.S.O., M.C. (*Adjt.*, Aux. Force, India) 26Nov.17
c.o. Mulholland, G. F. 25Dec.17

Lieutenants—contd.
2Harrison, H. B., M.C. 16Jan.18
d. 2Lawrence, L. A. 21Apr.18
2Shepard, D. O A. 27Apr.18
2Dolan, J. J., M. C. 26May19
2Flewett, T. E. 21June19
2Chadwick, T. A. H. 21June19
2Norman, D. S. 21June19
2Condron, M. A. 6July19
1Watson, L. R. C. 20Dec.20
2FitzGerald, M. H. 20Dec.20
2Russell, G. D. B. 30Dec.20
1Harrison, J. H. S. 16July21
1Renny, P. L. G. 17Dec.21
1Eassie, W. J. F. 17Dec.21

2nd Lieutenants.
1Hogan, K. E 16July20
1Rooth, R. G. 16July20
1Tarleton, B. 22Dec.21
1Houghton, H. W. H. 23Dec.21

Adjutants.
1Carroll-Leahy, T. J., D.S.O., M.C., *capt.* 15Oct.19
2Stitt, W. H., D.S.O., M.C, U. 1Nov.19

Quarter-Masters.
m. Holloway, L. 28Nov.08, *maj.* 1July17
1Williams, A. R. 8Mar.13, *capt.* 1July17
Dowling, J. E. 25Aug.14, *capt.* 25Aug.17 / 15Oct.21

[Appendix Ten

DISTRIBUTION OF OFFICERS' AND SERGEANTS' MESS PROPERTY.

COMMITTEE OF ADJUSTMENT, ROYAL DUBLIN FUSILIERS.

Meeting held at the Army and Navy Club, on Friday, June 9th, 1922.
The following were present :—

    Colonel Elford Pearse.
    Colonel Commandant H. W. Higginson, C.B., D.S.O., A.D.C.
    Lieut.-Colonel C. N. Perreau, C.M.G.
    Lieut.-Colonel G. S. Higginson.
    Major D. French.
    Major R. M. Watson, D.S.O.

Major-General C. F. Romer, C.B., C.M.G., was unable to attend.

IT WAS DECIDED :—

1. That certain articles as shown in the Inventory, should be sent to the Royal United Service Institution.

2. That articles as shown on the Inventories presented by past and present officers and other persons should be returned to the donors, or their heirs and successors for safe custody.

2A. In the case of articles presented by more than one person, such articles will be disposed of by drawing lots amongst those persons who are living at the time.

In the event of there being no survivors amongst the donors the Commanding Officers and sub-committees of Battalions shall have discretionary powers of disposal.

3. That the articles as shown in the Inventories (*i.e.* those won by the Battalions and those not presented by individuals) should be distributed among those officers of the 1st and 2nd Battalions who were members of the Royal Dublin Fusiliers Dinner Club in 1922, and who are willing to undertake the safe custody of these.

4. That miscellaneous articles, viz., furniture, books, pictures, etc., should be disposed of under instructions to be issued by the Commanding Officers and their sub-committees of the 1st and 2nd Battalions respectively.

5. That the silver forks and spoons should be stored for three years, and in the event of the Regiment not being reconstituted as laid down in para. 6, these should be distributed amongst those officers, or their heirs and successors, serving on the strength of the Regiment on the date of disbandment.

6. That the articles dealt with under paras. 2 and 3, should be distributed on the understanding that they shall be returned to the Regiment in the event of its being reconstituted as a Regiment of the Imperial British Army ; *i.e.* that it must form part of the British Army *as it now does*, and is not part of any Army of the Irish Free State.

7. That if the Regiment is not reconstituted as laid down in para. 6, within three years of the date of disbandment, the articles enumerated in paras. 2 and 3, should become the property of the persons in whose custody they were placed.

8. That all articles not claimed within one month of the date of the posting of the notification relating thereto, will be redistributed by the Committee.

9. That power be given to Commanding Officers and sub-committees of the 1st and 2nd Battalions to make presentations of plate to :—

(1) H.M. The King.

(2) Field-Marshal H.R.H. The Duke of Connaught, K.G., Colonel-in-Chief, The Royal Dublin Fusiliers.

(3) R. A. Bacon, Esq., O.B.E.

(4) The Committee, Royal Dublin Fusiliers Prisoners of War Fund, viz., Lady Arnott, Sir Stanley Cochrane, Baronet, and J. Whiteside Dane, Esq.

DISPOSAL OF THE SILVER AND MESS PROPERTY
2nd Bn. The Royal Dublin Fusiliers.

The Battalion Colours and a Facsimile of the " Tapp " Centrepiece have been sent to H.M. The King.

The two seven-light Candelabra have been sent to Field-Marshal H.R.H. The Duke of Connaught, K.G., etc.

The following articles have been sent to the Royal United Service Institution :—

    " Tapp " Centrepiece.
    Two Smaller Centrepieces.
    Forty-six Medals.
    One ten-branch Candelabra.
    Sikh Colours and Trophies.

The remainder of the silver if presented has been returned to the donor, and if not presented distributed amongst the members of the Royal Dublin Fusiliers Dinner Club.

BADGES AND BUTTONS

FULL DRESS— (1) Busby Grenade   (2) Tunic Grenades   (3) Tunic Buttons

SERVICE DRESS— (4) Cap Badge   (5) Collar Grenades   (6) Buttons

BLUE UNDRESS – (7) Collar Grenades

MESS DRESS— (8) Lapel Grenades   (9) Buttons for Jacket and Waistcoat

(10) SHOULDER BADGE FOR RANK AND FILE

[*Appendix Twelve*

CEREMONIAL OF THE RECEPTION OF THE COLOURS OF THE DISBANDED SOUTHERN IRISH REGIMENTS BY THE KING at Windsor Castle on Monday, the 12th June, 1922, at 11.30 a.m.

On Monday, the 12th June, 1922, at 11.30 a.m., in St. George's Hall, Windsor Castle, the King will take over the Colours of the following Regiments :—

> The Royal Irish Regiment,
> The Connaught Rangers,
> The Prince of Wales's Leinster Regiment (Royal Canadians),
> The Royal Munster Fusiliers, and
> The Royal Dublin Fusiliers.

His Majesty will also receive a Regimental Engraving offered by the South Irish Horse.

The detachments from the six Regiments will travel by the 9.55 a.m. train from Paddington Railway Station, arriving at Windsor G.W. Railway Station at 10.42 a.m.

On arrival at Windsor Railway Station the Colour Parties will proceed to the Royal Waiting Room, where the Colours will be unfurled.

The detachments will be met at Windsor Railway Station by an Escorting Party of 100 all ranks of the Third Battalion, Grenadier Guards, accompanied by the Band of the Regiment.

The Escorting Party will Present Arms as the five Colour Parties march out of the Royal Waiting Room with the Colours.

The six detachments of the Irish Regiments will be formed up in the following order, *i.e.*:—

> The Royal Irish Regiment,
> The Connaught Rangers,
> The South Irish Horse,
> The Prince of Wales's Leinster Regiment (Royal Canadians),
> The Royal Munster Fusiliers,
> The Royal Dublin Fusiliers.

The detachments, headed by the Band and one-half the Escorting Party, the other half bringing up the rear of the Column, will march to the Quadrangle of the Castle.

At the gateway of the Castle, the Troops will be met by Lieutenant-Colonel The Marquis of Cambridge, Governor and Constable of Windsor Castle.

On arrival in the Quadrangle of Windsor Castle, the six detachments of the Irish Regiments will enter by the Grand Entrance and proceed to St. George's Hall, the Escorting Party, with Band, remaining in the Quadrangle.

The six detachments of the Irish Regiments will form up in St. George's Hall facing the windows, in the following order, numbering from the right :—

1. The Royal Irish Regiment,
2. The Connaught Rangers,
3. The South Irish Horse,
4. The Prince of Wales's Leinster Regiment (Royal Canadians),
5. The Royal Munster Fusiliers,
6. The Royal Dublin Fusiliers.

The King, accompanied by Field-Marshal His Royal Highness The Duke of Connaught (Honorary Colonel of the South Irish Horse and Colonel-in-Chief of the Royal Dublin Fusiliers), will enter St. George's Hall at 11.30 a.m. Upon His Majesty taking up his position facing the centre of the line, a Royal Salute will be given, the Colours will be lowered, and the Band in the Quadrangle will play the National Anthem.

The King will then inspect the six detachments and resume his place facing the centre of the line.

His Majesty will address the detachments ; after which the Ceremony of Presentation of the Colours will take place, concluding with the offering by the South Irish Horse of the Engraving.

On the conclusion of the Ceremony a Royal Salute will be given, the Band in the Quadrangle playing the National Anthem.

His Majesty leaves St. George's Hall.

Dress.   Service Dress.
         Grenadier Guards. Full Dress.
         Detachments from Irish Regiments wear Medals and Breast
            Decorations.

COMPOSITION OF THE SIX DETACHMENTS FROM IRISH REGIMENTS :—

THE SOUTH IRISH HORSE.

Honorary Colonel	...	F.M. H.R.H. The Duke of Connaught.
Colonel ...	...	Lieutenant-Colonel I. W. Burns-Lindow, D.S.O.
Second in Command	...	Major R. Smyth.
Adjutant ...	...	Captain R. Dease.
		Sergeant Goodchild.

THE ROYAL IRISH REGIMENT.

Colonel ...	...	Major-General J. Burton Forster, C.B.
O.C. 1st Battn.	...	Major F. Call, D.S.O.
King's Colour ...	...	Lieutenant J. J. Burke-Gaffney, M.C.

Regimental Colour	...	Lieutenant W. C. V. Galwey, M.C.
		R.S.M. C. Field.
		C.S.M. J. Bridger, M.S.M.
		Sergeant H. Taylor, M.M.
		Sergeant W. Dunne, D.C.M.
O.C. 2nd Battn.	...	Lieutenant-Colonel G. A. Elliot, M.C.
King's Colour ...	...	Lieutenant W. C. L. Shee.
Regimental Colour	...	Lieutenant E. C. Beard, M.C.
		R.S.M. R. Burns.
		C.S.M. J. Cussens, D.C.M.
		C.S.M. J. Bergin.
		C.S.M. B. Harris.

THE CONNAUGHT RANGERS.

O.C. 1st Battn.	...	Lieutenant-Colonel W. N. S. Alexander, D.S.O.
		Captain F. D. Foott.
		Captain W. O'Brien. M.C.
		Sergeant Wallace.
		Sergeant Malone.
		Sergeant Scott.
O.C. 2nd Battn.	...	Lieutenant-Colonel H. F. N. Jourdain, C.M.G.
		Lieutenant G. B. Champion.
		Lieutenant C. G. Gaden, M.C.
		R.S.M. M. J. Monaghan, M.C.
		Bandmaster G. Landrock.
		Acting R.S.M. P. J. Finucane, D.C.M.

THE PRINCE OF WALES'S LEINSTER REGIMENT (ROYAL CANADIANS).

Colonel ...	...	Major-General G. F. Boyd, C.B., C.M.G., D.S.O.
O.C. 1st Battn.	...	Colonel E. T. Humphreys, C.M.G., D.S.O.
		Captain T. B. Deane.
		Lieutenant F. A. Levis.
		C.S.M. A. Bradley, M.M.
		C.S.M. J. Newton.
		C.Q.M.S. A. Madden.
O.C. 2nd Battn.	...	Lieutenant-Colonel R. A. H. Orpen-Palmer, D.S.O.
		Captain W. S. Caulfeild, M.C.
		Captain T. E. M. Battersby.
		R.S.M. C. H. Smith, M.C., D.C.M.
		C.S.M. J. Finn.
		Colour-Sergeant J. Cannon.

THE ROYAL MUNSTER FUSILIERS.

Colonel	... ...	Lieutenant-General Sir H. S. G. Miles, G.C.B., G.C.M.G., G.B.E., C.V.O.
O.C. 1st Battn.	...	Lieutenant-Colonel J. A. F. Cuffe, C.M.G., D.S.O.
		Major G. W. Geddes, D.S.O.
		Captain G. R. Prendergast.
		C.Q.M.S. Fitzmaurice.
		Private McNamara.
		Private Wynne.
O.C. 2nd Battn.	...	Lieutenant-Colonel H. S. Jervis, M.C.
King's Colour ...	...	Major C. R. Rawlinson.
Regimental Colour	...	Lieutenant and Adjutant C. B. Callander, M.C.
		R.S.M. J. Ring, M.C., D.C.M.
		Lance-Corporal J. Foley.
		Private J. Merner, M.M.

THE ROYAL DUBLIN FUSILIERS.

Colonel-in-Chief	...	F.M. H.R.H. The Duke of Connaught.
O.C. 1st Battn.	...	Lieutenant-Colonel C. N. Perreau, C.M.G.
King's Colour	...	Major J. P. Tredennick, D.S.O., O.B.E.
Regimental Colour	...	Major T. J. Carroll-Leahy, D.S.O., M.C.
		C.S.M. A. Cullen.
		Sergeant T. Doyle.
		Sergeant A. D. Connolly.
O.C. 2nd Battn.	...	Lieutenant-Colonel G. S. Higginson.
King's Colour	...	Captain J. M. Mood, O.B.E., M.C.
Regimental Colour	...	Captain C. G. Carruthers, M.C.
		Colour-Sergeant J. A. Jones.
		Colour-Sergeant P. Kehoe.
		Colour-Sergeant G. Sexton.

[*Note.*—Major-General C. D. Cooper, C.B., Colonel of The Royal Dublin Fusiliers, was unavoidably prevented from being present.]

[Appendix Thirteen

THE ROYAL DUBLIN FUSILIERS ASSOCIATION.

(FORMED IN 1910.)

Patron : Field-Marshal H.R.H. THE DUKE OF CONNAUGHT, K.G., K.T., K.P., G.C.B., G.C.S.I., G.C.M.G., G.C.I.E., G.C.V.O., G.B.E., Colonel-in-Chief The Royal Dublin Fusiliers.

President : Major-General C. D. COOPER, C.B., Colonel of the Regiment.

Vice-President : Brigadier-General G. DOWNING.

Hon. Treasurer : Lieut.-Colonel E. ST. G. SMITH.

Secretary : Major R. BAKER, D.S.O.

Hon. Medical Adviser : Colonel Sir W. T. de C. WHEELER, F.R.C.S., 23, Fitzwilliam Square, Dublin.

Temporary Office : 77, Grosvenor Square, Rathmines, Dublin.

The Objects of the Association are—

1. To maintain connection between the men serving in The Royal Dublin Fusiliers and old comrades, and to promote friendship and association amongst those now in civilian life interested in their old Regiment.

2. To promote the welfare of discharged soldiers of the Royal Dublin Fusiliers who are members of this Association, by assisting them to obtain situations, and generally to aid members in establishing themselves respectably in civilian life.

3. To assist financially any member who, through no fault of his own, has fallen into bad health, and is unable to earn his own livelihood or who is out of employment and in distressed circumstances, if the Committee to whom application is made deem the applicant worthy ; and also to assist financially widows or children of members who are in distressed circumstances.

4. In case of death to provide for the burial of any member, if aid is required.

5. To foster *esprit de corps* and promote recruiting for the Regiment, and to make widely known the advantages of service in the Regiment, and to create a brotherhood between the non-commissioned officers and men after discharge from either Battalion.

6. To make application to the authorities on behalf of veterans for pensions where none have already been granted, or for increased allowance where only small sums are received.

Keeping the above objects in view, any member of the Association should inform the Committee of any suitable situation which he may know to be vacant.

Members who are desirous of employment should make their wants known to the Secretary.

At the Meeting of the Past and Present Officers of the Regiment, held in London on the 9th March, 1922, it was decided to continue the Regimental Association after the disbandment of the Regiment.

It is expected that in the near future applications for assistance will be much increased owing to so many men, now in military employment in Ireland, losing their employment as a result of the withdrawal of troops from the country.

There are many men, most of them still serving, who, having paid for from four to ten years membership after discharge, will have a claim on the Association until the expiration of these periods.

In these circumstances it is hoped that all Officers will continue their subscriptions. The rate of Annual Subscription is 10/6 for Subalterns and 21/- for higher ranks. Subscriptions may be paid through Messrs. Cox & Co., or sent direct to the Secretary, Major Baker, D.S.O., 77, Grosvenor Square, Rathmines, Co. Dublin.

[*Appendix Fourteen*

REGIMENTAL SONG

(Common to both Battalions).

I

COME, boys, we'll sing of gallant lads,
  " Old Toughs " of bygone days ;
A roll of names we claim our own,
  And wreathe with choicest bays ;
Of days ere Britain's name was great
  In India's burning zone,
Yet shone with brightness guarded there
  By our " Old Toughs " alone.

CHORUS.

" Old Toughs " were they, " Old Toughs " are we !
  And through the coming years
We'll keep that name without a stain
  As " Dublin Fusiliers."

II

Two hundred years and more, my boys,
  Since first was marched away
From Britain's shore our fine old corps,
  To keep for her—Bombay ;
And there, when Englishmen were few
  And foes were strong around,
Their only safeguard many a year
  In our " Old Toughs " was found.

Chorus—" Old Toughs " were they, etc.

III

As time sped on and gave command
  For Britain's sway to spread,
Bombay sent forth her pioneers,
  Who in the vanguard led ;
They led a hundred fights, my lad !

They stormed the strongest wall !
And Irishmen were there, who shed
  Their heart's best blood through all.
Chorus—" Old Toughs " were they, etc.

IV

No tale of ancient chivalry,
  No mediæval lays,
Can tell of deeds more wonderful
  Than since those early days,
When, first as " Bombay Regiment,"
  Then " Bombay Fusiliers,"
They wrapt with glory Britain's name
  Through twice a hundred years.
Chorus.—" Old Toughs " were they, etc.

V

Our Colours, boys, but little tell
  The fame of our old corps ;
A dozen flags we well could fill,
  And maybe even more ;
Through length and breadth of Hindostan,
  Bombay, Madras, Bengal,
From Plassy, east, to Aden, west,
  " Old Toughs " is stamped on all.
Chorus.—" Old Toughs " were they, etc.

VI

Some names we never should forget,
  Of men, in days gone by,
Who went, with " hope forlorn," through death,
  To plant our standard high ;
At Ahmedabad, like Hieme and Fridge,
  Who led a dauntless few ;
At S'ringapatam, where Graham died
  At Mooltan, Bennett too.
Chorus—" Old Toughs " were they, etc.

VII

The hearts at home throbbed warm, my lads,
  When those brave deeds were told,
Of heroes we may call our comrades ;
  Should our hearts grow cold ?
From Arcot down to Mut'ny days,
  From Havelock back to Clive,
The " Toughs " have won a glorious name,
  'Tis ours to keep alive.

Chorus—" Old Toughs " were they, etc.

VIII

Should we be ever called to share
  In scenes like those of yore,
We know the stuff that's in us, lads,
  Will bring us to the fore.
But times of peace must try our pluck
  With test both close and keen ;
For honour, boys, then keep our name,
  Like shamrocks, fair and green.

Chorus—" Old Toughs " were they, etc.

MELODY OF REGIMENTAL SONG.

INDEX

                                          R